From Clocks to Chaos

From Clocks
to Chaos

The Rhythms
of Life

Leon Glass and
Michael C. Mackey

Princeton
University
Press

Published by Princeton University Press, 41 William Street, Princeton,
New Jersey 08540
In the United Kingdom: Princeton University Press, Guildford, Surrey

This book has been composed in Linotron Times Roman by Syntax International
Clothbound editions of Princeton University Press books are printed on
acid-free paper, and binding materials are chosen for strength and durability.
Paperbacks, although satisfactory for personal collections,
are not usually suitable for library rebinding

Printed in the United States of America by Princeton University Press,
Princeton, New Jersey

Designed by Laury A. Egan

Library of Congress Cataloging-in-Publication Data

Glass, Leon 1943–
From clocks to chaos.

Bibliography: p.   Includes index.
1. Biological rhythms.   2. Biological rhythms—
Mathematics.   I. Mackey, Michael C., 1942–
II. Title.
QH527.G595 1988   574.1′882′0151   87-32803
ISBN 0-691-08495-5 (alk. paper)
ISBN 0-691-08496-3 (pbk.)

To Our Children

Contents

This book deals with the applications of mathematics to the study of normal and pathological physiological rhythms. It is directed toward an audience of biological scientists, physicians, physical scientists, and mathematicians who wish to read about biological rhythms from a theoretical perspective.

Throughout this volume, we discuss many biological examples and present selected mathematical models to emphasize main concepts. The biological examples have been chosen to illustrate the great variety of dynamic processes occurring in different organ systems. For most of the biological examples, a definitive theoretical interpretation is impossible at the current time. Consequently, the mathematical models are not intended to be exact descriptions of the real biological system, but are simplified approximations. We have tried to emphasize the main principles and to present them in the simplest way possible. It will remain for future researchers to determine whether more realistic models display the same dynamical properties as the simplified versions we present.

We assume a knowledge of calculus but try to explain all advanced concepts and intend the text to be intelligible to nonmathematicians. Equations are used sparingly, and we illustrate ideas with physiological examples and graphs whenever possible. Although there are frequent cross-references between chapters, the chapters are largely independent of one another and do not have to be read in the sequence presented. However, readers with little background in mathematics will need to refer back to chapters 2 and 3 for explanations of unfamiliar concepts. The Mathematical Appendix gives further details of some of the main mathematical techniques together with examples and problems to illustrate the application of these techniques in concrete situations.

Because of the large range of potential applications of the theory, it has been impossible to give exhaustive references. Rather, we have tried to give several key references for each topic to assist the reader in identifying the relevant literature. In order to preserve the flow of the text, we have collected the references in the separate Notes and References sections that follow each chapter.

Over the years we have benefited enormously from discussions and collaborations with students and colleagues. In particular we thank J. Bélair, P. Dörmer, A. Goldberger C. Graves, M. R. Guevara, U. an der Heiden, S. A. Kauffman, J. Keener, A. Lasota, J. G. Milton,

R. Perez, G. A. Petrillo, A. Shrier, T. Trippenbach, and A. T. Winfree. J. G. Milton, S. Strogatz, J. Tyson, and A. T. Winfree made many useful suggestions concerning presentation of the text, and J. G. Milton suggested the main title. The figures were drafted by B. Gavin, and S. James helped with the typing. We would like to thank Judith May and Alice Calaprice of Princeton University Press for their help and advice throughout the production of this book.

This book was partially written while LG was a visiting research scientist at the University of California at San Diego and MCM was a visiting professor at the Universities of Oxford and Bremen. We thank H. Abarbanel and A. Mandell (San Diego), J. D. Murray (Oxford), and H. Schwegler (Bremen) for their hospitality during this period of time. Finally, we have benefited from research grants from the Natural Sciences and Engineering Research Council (Canada), the Canadian Heart Association, and the Canadian Lung Foundation.

Montreal
August 1987

Sources and Credits

Sources of previously published figures are acknowledged in the captions. Additional information is listed below. In some cases figures have been modified to improve legibility. Our thanks to the authors and publishers for permission to reproduce these figures.

Figure
1.1   Hosomi, H., and Hayashida, Y. 1984. Systems analysis of blood pressure oscillation. In *Mechanisms of Blood Pressure Waves*, ed. K. Miyakawa, H. P. Koepchen, and C. Polosa, pp. 215–27.
1.2, 8.2a   Goldberger, A. L., and Goldberger, E. 1986. *Clinical Electrocardiography: A Simplified Approach*, ed. 3. St Louis: C. V. Mosby.
1.3   Molnar, G. D., Taylor, W. F., and Langworthy, A. L. 1972. Plasma immunoreactive insulin patterns in insulin-treated diabetics. *Mayo Clin. Proc.* 47: 709–19.
1.4   Kiloh, L. G., McComas, A. J., Osselton, J. W., and Upton, A.R.M. 1981. *Clinical Electroencephalography*. London: Butterworths.
1.5   Sakmann, B., Noma, A., and Trautwein, W. 1983. Acetylcholine activation of single muscarinic K^+ channels in isolated pacemaker cells of the mammalian heart. *Nature* 303: 250–53. *Copyright* © 1983 Macmillan Magazines Limited.
1.8, 4.9   Mackey, M. C., and Glass, L. 1977. Oscillation and chaos in physiological control systems. *Science* 197: 287–89. Copyright 1977 by the AAAS.
1.9   Jalife, J., and Antzelevitch, C. 1979. Phase resetting and annihilation of pacemaker activity in cardiac tissue. *Science* 206: 695–97. *Copyright* 1979 *by the AAAS.*
1.10, 2.2b   Glass, L., Guevara, M. R., Bélair, J., and Shrier, A. 1984. Global bifurcations of a periodically forced biological oscillator. *Phys. Rev.* 29: 1348–57.
1.11, 7.11   Glass, L., Shrier, A., and Bélair, J. 1986. Chaotic cardiac rhythms. In *Chaos*, ed. A. Holden, pp. 237–56. Manchester: Manchester University Press.
1.12   Figure provided by A. T. Winfree.
2.2a   Figure provided by A. Shrier.
2.6   Figure provided by J. Crutchfield.
3.1   Fatt, P., and Katz, B. 1952. Spontaneous subthreshold activity at motor nerve endings. *J. Physiol. (Lond.)*. 117: 109–28.

3.2a   Rodieck, R. W., Kiang, N. Y.-S., and Gerstein, G. 1962. Some quantitative methods for the study of spontaneous activity of single neurons. *Biophys. J.* 2: 351–68. By copyright permission of the Biophysical Society.

3.2b, 3.4   Gerstein, G. L., and Mandelbrot, M. 1964. Random walk models for the spike activity of a single neuron. *Biophys. J.* 4: 41–68. By copyright permission of the Biophysical Society.

3.3   Lasota, A., and Mackey, M. C. 1985. *Probabilistic Properties of Deterministic Systems.* Cambridge Eng.: Cambridge University Press.

3.10   Aihara, K., Numajiri, T., Matsumoto, G., and Kotani, M. 1986. Structures of attractors in periodically forced neural oscillators. *Phys. Lett. A* 116: 313–17.

3.11   Mandelbrot, B. B. 1982. *The Fractal Geometry of Nature.* San Francisco: W. H. Freeman.

4.1   Hodgkin, A. L., and Huxley, A. F. 1952. A quantitative description of membrane current and its application to conduction and excitation in nerve. *J. Physiol. (Lond.)* 117: 500–44.

4.2a   McAllister, R. E., Noble, D., and Tsien, R. W. 1975. Reconstruction of the electrical activity of cardiac Purkinje fibers. *J. Physiol. (Lond.)* 251: 1–59.

4.2b   Noble, D. 1984. The surprising heart: A review of recent progress in cardiac electrophysiology. *J. Physiol. (Lond.)* 353: 1–50.

4.3   Lebrun, P., and Atwater, I. 1985. Chaotic and irregular bursting of electrical activity in mouse pancreatic β-cells. *Biophys. J.* 48: 529–31. By copyright permission of the Biophysical Society.

4.4   Chay, T. R., and Rinzel, J. 1985. Bursting, beating and chaos in an excitable membrane model. *Biophys. J.* 45: 357–66. By copyright permission of the Biophysical Society.

4.6   Selverston, A. I., Miller, J. P., and Wadepuhl, M. 1983. Cooperative mechanisms for the production of rhythmic movements. In *Neural Origin of Rhythmic Movements,* ed. A. Roberts and B. Roberts, pp. 55–87. Soc. Exp. Biol. Symposium 37.

4.7   Cohen, M. I. 1974. The genesis of respiratory rhythmicity. In *Central-Rhythmic and Regulation,* ed. W. Umbach and H. P. Koepchen, pp. 15–35. Stuttgart, W. Germany: Hippokrates.

4.8   Glass, L., and Young, R. E. 1979. Structure and dynamics of neural network oscillators. *Brain Res.* 179: 207—18.

4.10   Stark, L. W. 1968. *Neurological Control Systems: Studies in Bioengineering.* New York: Plenum.

4.13   Mackey, M. C., and an der Heiden, U. 1984. The dynamics of recurrent inhibition. *J. Math. Biol.* 19: 211–25.

5.1   Bortoff, A. 1961. Electrical activity of intestine recorded with pressure electrode. *Am. J. Physiol.* 201: 209–12.

5.2   Guevara, M. R. 1987. Afterpotentials and pacemaker oscillations in an ionic model of cardiac Purkinje fibre. In *Temporal Disorder in Human Oscillatory Systems*, ed. L. Rensing, U. an der Heiden, and M. C. Mackey, pp. 126–33. Berlin: Springer-Verlag.

5.4   Schulman, H., Duvivier, R., and Blattner, P. 1983. The uterine contractility index. *Am. J. Obstet. Gynecol.* 145: 1049–58.

5.5, 5.6   Glass, L. 1987. Is the respiratory rhythm generated by a limit cycle oscillation? In *Concepts and Formalizations in the Control of Breathing*, ed. G. Benchetrit, P. Baconnier, and J. Demongeot, pp. 247–63. Manchester: Manchester University Press.

5.7   McClellan, A. D., and Grillner, S. 1984. Activation of 'fictive swimming' by electrical microstimulation of brainstem locomotor regions in an *in vitro* preparation of the lamprey central nervous system. *Brain Res.* 300: 357–61.

5.9   Guttman, R., Lewis, S., and Rinzel, J. 1980. Control of repetitive firing in squid axon membrane as a model for a neuron oscillator. *J. Physiol. (Lond.)* 305: 377–95.

5.10   Petersen, I., and Stener, I. 1970. An electromyographical study of the striated urethral sphincter, the striated anal sphincter, and the levator ani muscle during ejaculation. *Electromyography* 10: 24–44.

6.1a   Clark, F. J., and Euler, C. von 1972. On the regulation of depth and rate of breathing. *J. Physiol. (Lond.)* 222: 267–95.

6.1b   Knox, C. K. 1973. Characteristics of inflation and deflation reflexes during expiration in the cat. *J. Neurophysiol.* 36: 284–95.

6.2   Jalife, J., and Moe, G. K. 1976. Effect of electrotonic potential on pacemaker activity of canine Purkinje fibers in relation to parasystole. *Circ. Res.* 39: 801–808. By permission of the American Heart Assoc., Inc.

6.7   Glass, L., and Winfree, A. T. 1984. Discontinuities in phase-resetting experiments. *Am. J. Physiol.* 246 (*Regulatory Integrative Comp. Physiol. 15*): R251–58.

6.8   Lund, J. P., Rossignol, S., and Murakami, T. 1981. Interactions between the jaw opening reflex and mastication. *Can. J. Physiol. Pharmacol.* 59: 683–90.

6.9   Stein, R. B., Lee, R. G., and Nichols, T. R. 1978. Modifications of ongoing tremors and locomotion by sensory feedback. *Electroencephalogr. Clin. Neurophysiol.* (Suppl.) 34: 511–19.

6.10   Castellanos, A., Luceri, R. M., Moleiro, F., Kayden, D. S., Trohman, R. G., Zaman, L., and Myerburg, R. J. 1984. Annihilation, entrainment and modulation of ventricular parasystolic rhythms. *Am. J. Cardiol.* 54: 317–22.

6.11   Guevara, M. R., Shrier, A., and Glass, L. 1986. Phase resetting of spontaneously beating embryonic ventricular heart-cell aggregates. *Amer. J. Physiol.* 251 (*Heart Circ. Physiol. 20*): H1298–H1305.

7.1, 7.8   Petrillo, G. A., and Glass, L. 1984. A theory for phase locking of respiration in cats to a mechanical ventilator. *Am. J. Physiol.* 246 (*Regulatory Integrative Comp. Physiol. 15*): R311–20.

7.2, 8.4b, c, d   Glass, L., Guevara, M. R., and Shrier, A. 1987. Universal bifurcations and the classification of cardiac arrhythmias. *Ann. N. Y. Acad. Sci.* 504: 168–178.

7.3   Hayashi, C. 1964. *Nonlinear Oscillations in Physical Systems.* New York: McGraw Hill.

7.6   Glass, L., and Mackey, M. C. 1979b. A simple model for phase locking of biological oscillators. *J. Math. Biol.* 7: 339–52.

7.7, 7.10   Glass, L., and Bélair, J. 1986. Continuation of Arnold tongues in mathematical models of periodically forced biological oscillators. In *Nonlinear Oscillations in Biology and Chemistry*, ed. H. G. Othmer, pp. 232–43. Berlin: Springer-Verlag.

7.12   Bramble, D. M. 1983. Respiratory patterns and control during unrestrained human running. In *Modelling and Control of Breathing*, ed. B. J. Whipp and D. M. Wiberg, pp. 213–20. New York: Elsevier. Copyright 1983 Elsevier Science Publishing Co, Inc.

7.13   Moe, G. K., Jalife, J., Mueller, W. J., and Moe, B. 1977. A mathematical model of parasystole and its application to clinical arrhythmias. *Circulation* 56: 968–79. By permission of the American Heart Association, Inc.

7.14, 7.15   Graves, C., Glass, L., Laporta, D., Meloche, R., and Grassino, A. 1986. Respiratory phase locking during mechanical ventilation in anesthetized human subjects. *Am. J. Physiol.* 250 (*Regulatory Integrative Comp. Physiol. 19*): R902–R909.

8.1a   Weiss, R. M., Wagner, M. L., and Hoffman, B. F. 1968. Wenckebach periods of the ureter: A further note on the ubiquity of the Wenckebach phenomenon. *Invest. Urol.* 5: 462–67. © by Williams and Wilkins, 1968.

8.1b   Prosser, C. L., Smith, C. E., and Melton, C. E. 1955. Conduction of action potentials in the ureter of the rat. *Am. J. Physiol.* 181: 651–60.

8.2b   Bellett, S. 1971. *Clinical Disorders of the Heartbeat.* Philadelphia: Lea & Febiger.

8.4a   Levy, M. N., Martin, P. J., Edelstein, J., and Goldberg, L. B.

1974. The AV nodal Wenckebach phenomenon as a positive feedback mechanism. *Prog. Cardiovasc. Dis.* 16: 601–13.

8.5   Guevara, M. R., Ward, G., Shrier, A., and Glass, L. 1984. Electrical alternans and period-doubling bifurcations. In *Computers in Cardiology*, pp. 167–70. © 1984 IEEE.

8.6   Sarna, S. K. 1985. Cyclic motor activity: Migrating motor complex. *Gastroenterology* 89: 894–913. Copyright 1985 by the American Gastroenterological Association.

8.8   Winfree, A. T. 1973. Scroll-shaped waves of chemical activity in three dimensions. *Science* 181: 937–39. Copyright 1973 by the AAAS.

8.9a   Winfree, A. T., and Strogatz, S. H. 1984. Organizing centers for three-dimensional chemical waves. *Nature* 311: 611–15. Copyright © 1984 Macmillan Magazines Limited.

8.9b   Welsh, B., Gomatam, J., and Burgess, A. E. 1983. Three-dimensional chemical waves in the Belousov-Zhabotinsky reaction. *Nature* 304: 611–14. Copyright © 1983 Macmillan Magazines Limited.

8.10   Downar, E., Parson, I. D., Mickleborough, L. L, Cameron, D. A., Yao, L. C., and Waxman, M. B. 1984. On-line epicardial mapping of intraoperative ventricular arrhythmias: Initial clinical experience *J. Amer. Coll. Cardiol.* 4: 703–14. Reprinted with permission from the American College of Cardiology.

8.11   Shibata, M., and Bures, J. 1972. Reverberation of cortical spreading depression along closed-loop pathways in rat cerebral cortex. *J. Neurophysiol.* 35: 381–88.

A.7   Glass, L., Guevara, M. R., Shrier, A., and Perez, R. 1983. Bifurcation and chaos in a periodically stimulated cardiac oscillator. *Physica* 7D: 89–101.

A.8   Bélair, J., and Glass, L. 1985. Universality and self-similarity in the bifurcations of circle maps. *Physica* 16D: 143–54.

From Clocks to Chaos

Chapter 1

Introduction:
The Rhythms of Life

$\mathbf{P}$hysiological rhythms are central to life. Some rhythms are maintained throughout life, and even a brief interruption leads to death. Other rhythms, some under conscious control and some not, make their appearance for various durations during an individual's life. The rhythms interact with one another and with the external environment. Variation of rhythms outside of normal limits, or appearance of new rhythms where none existed previously, is associated with disease.

An understanding of the mechanisms of physiological rhythms requires an approach that integrates mathematics and physiology. Of particular relevance is a branch of mathematics called nonlinear dynamics. The roots of nonlinear dynamics were set by Poincaré at the end of the last century but have seen remarkable developments over the past 25 years. Unfortunately, the main features of nonlinear dynamics are usually presented in a format suitable for advanced students in mathematics and are thus difficult for the practicing physiologist. Yet many of the central ideas that are most relevant in physiology can be expressed and illustrated in concrete physiological examples. This book is intended to offer an introduction to recent advances in nonlinear dynamics as they have been applied to physiology, in a format intelligible to a nonmathematician. However, we also hope that those with a mathematical background will find the numerous physiological examples of interest, and that some will even find the many poorly understood phenomena in physiology which we discuss a stimulus for future research. In this chapter we give a brief outline of this book and summarize its themes by giving several physiological examples.

1.1 Mathematical Concepts

It is common to measure physiological observables as a function of time. Four main mathematical ideas have been developed to characterize such time series: steady states, oscillations, chaos, and noise.

Since the pioneering research of Bernard, Cannon, and others, it has become fashionable, if not obligatory, to discuss homeostasis near the beginning of physiology texts. *Homeostasis* refers to the relative constancy of the internal environment with respect to variables such as blood sugar, blood gases, electrolytes, osmolarity, blood pressure, and pH. The physiological concept of homeostasis can be associated with the notion of steady states in mathematics. *Steady states* refer to a constant solution of a mathematical equation. Elucidation of the mechanisms that constrain variables to narrow limits constitutes a key area of physiological research. As an example of a homeostatic mechanism, consider the response to a quick mild hemorrhage in an anesthetized dog (figure 1.1). Following the hemorrhage, reflex mechanisms are activated which restore blood pressure to near equilibrium within a few seconds.

Although the mean blood pressure is maintained relatively constant, as we all know, the contractions of the heart are approximately periodic. The periodic electrical activity of the heart can be visualized using an electrocardiogram. Figure 1.2 shows an example of a normal electrocardiogram. Likewise, all of us are familiar with the rhythms of heartbeat, respiration, reproduction, and the normal sleep-wake cycle. Less obvious, but of equal physiological importance, are oscillations in numerous other systems—for example, release of insulin and luteinizing hormone, peristaltic waves in the intestine and ureters, electrical activity of the cortex and autonomic nervous system, and constrictions in peripheral blood vessels and the pupil. Physiological oscillations are associated with periodic solutions of mathematical equations.

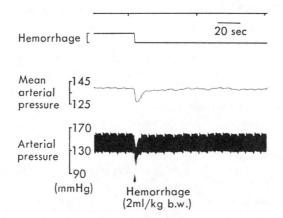

1.1. Arterial and mean arterial pressure responses to a quick mild hemorrhage in a dog anesthetized with sodium pentobarbital. From Hosomi and Hayashida (1984).

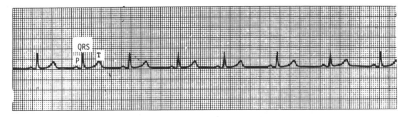

1.2. Normal electrocardiogram. The P wave corresponds to atrial depolarization, the QRS complex to ventricular depolarization, and the T wave to ventricular repolarization. One large box corresponds to 0.2 sec in the horizontal direction, and 0.5 mv in the vertical direction. From Goldberger and Goldberger (1986).

Of course, we all know that close measurement of any physiological variable will never give a time sequence that is absolutely stationary or periodic. Even systems that are assumed to be stationary or periodic will always have fluctuations about the fixed level or periodic cycle. In addition, there are systems that appear to be so irregular that it may be difficult to associate them with any underlying stationary or periodic process. One potential source of physiological variability is the fluctuating environment. As one eats, exercises, and rests, blood-sugar levels and insulin levels respond in a characteristic fashion (figure 1.3).

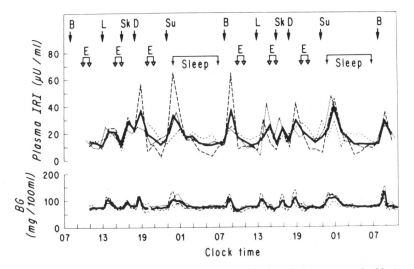

1.3. Immunoreactive insulin (IRI) and blood glucose (BG) in ambulatory normal subjects over a 48-hour period. Interrupted lines describe patterns for individual subjects; continuous lines show the group averages. Symbols: B = breakfast; L = lunch; Sk = snack; D = dinner; Su = supper; E = 1 hour of walking exercise. From Molnar, Taylor, and Langworthy (1972).

Similarly, the blood pressure responds to the changes in activity and posture. Physiological rhythms themselves can also act to perturb other rhythms. An example is respiratory sinus arrhythmia in which the heartbeat is quickened during inspiration. Although such variability is not necessarily easy to deal with theoretically, its origin is often readily understood.

More mysterious situations are those in which fluctuations are found even when environmental parameters are maintained at as constant a level as possible and no perturbing influences can be identified. For example, the electroencephalogram measures average electrical activity from the localized regions of the cortex and shows fluctuations over time which are often quite irregular (figure 1.4). These situations afford significant difficulties in understanding the mechanisms leading to the irregularities.

Mathematics offers us two distinct ways to think about the irregularities intrinsic to physiology. The more common of the two is *noise*, which refers to chance fluctuations. For example, such chance fluctuations are often associated with the opening and closing of channels in neurons and cardiac cells that carry ionic current (figure 1.5). Although "chaos" is often used as a popular synonym for noise, it has developed a technical meaning that is quite different. Technically, chaos refers to randomness or irregularity that arises in a deterministic system. In other words, chaos is observed even in the complete absence of environmental noise. An important aspect of chaos is that there is a sensitive dependence of the dynamics to the initial conditions. This means that although in principle it should be possible to predict future

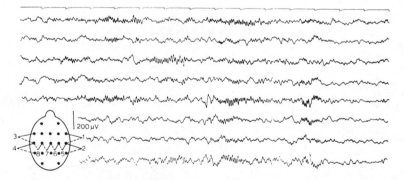

1.4. Electroencephalogram recorded from a normal 17-year-old woman during natural sleep. There are 14 Hz spindles, which are independent on either side. The top line shows 1-sec intervals. Simultaneous recordings from the eight electrode positions indicated on the diagram are displayed. From Kiloh et al. (1981).

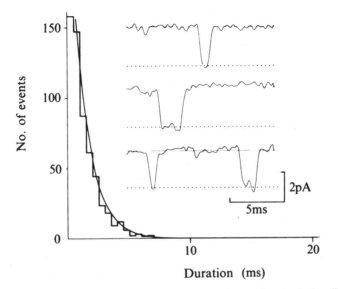

1.5. Currents flowing through an individual potassium channel from a single cell from dispersed AV node of rabbit heart. Short pulses of 2.4 pA amplitude at a resting potential of −20 mv. The histogram represents the distribution of current pulse durations and is fitted by a single exponential. From Sakmann, Noma, and Trautwein (1983).

dynamics as a function of time, this is in reality impossible since any error in specifying the initial condition, no matter how small, leads to an erroneous prediction at some future time.

Some equations display dynamics that are not periodic and fluctuate in irregular fashion. The existence of such equations was known to Poincaré and later mathematicians, but the recognition of these phenomena has only recently emerged in the natural sciences. The implications of such phenomena in biology and physiology are a topic of great current interest.

In practical situations, there are fluctuations about some mean value or oscillations which are more or less regular. It is not a trivial problem to go backwards from the observation of such dynamics to infer something about the underlying dynamical system.

Chapters 2 and 3 offer an introduction to the concepts of steady states, oscillations, noise, and chaos in mathematics. We show how these properties can arise in equations and how transitions between different types of dynamical behavior can occur. Since some of the material in chapters 2 and 3 is elementary, those with some knowledge of mathematics may wish to skip some of the sections. On the other hand, those with a weaker background in mathematics and those who really do not like

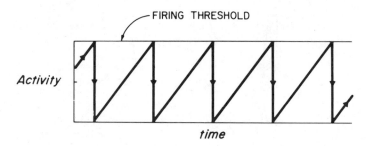

1.6. Integrate and fire model. The activity rises to a firing threshold and then resets to zero.

to read about mathematical ideas can skip ahead to other chapters, using chapters 2 and 3 as references as the need arises.

1.2 Mathematical Models for Biological Oscillators

There is a large literature that proposes many different types of models for the generation of physiological rhythms. The simplest type of model is called an *integrate and fire model*. In such models a quantity called the *activity* rises to a threshold leading to an event. The activity then instantaneously relaxes back to a second lower threshold. This process is represented schematically in figure 1.6. If the function determining the rise and fall of the activity between the two thresholds is fixed, and if the thresholds are fixed, then a periodic sequence of events will be generated at a readily determined frequency.

A physiological system that can be modeled by an integrate and fire mechanism is the one controlling the micturition reflex. As time proceeds, the bladder fills and eventually micturition takes place. Then the cycle starts anew. In the normal adult, micturition occurs 6–10 times/day with a voiding volume of 300–600 ml. However, pregnant women and patients with serious bladder or prostatic pathology often display increased frequency, reduced volume, and nocturia. In figure 1.7 we show the voided volume and micturition times recorded by a patient with carcinoma of the bladder. We are unaware of detailed quantitative studies or theoretical analysis of the micturition reflex or its pathological variants. A variety of other systems have been modeled by integrate and fire models, and we shall utilize such models in many different points in the text.

Although integrate and fire models are frequently used in physiology and will be discussed in several subsequent chapters, from a mathe-

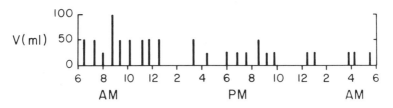

1.7. Micturition times and voided volume recorded by a patient with carcinoma of the bladder. Data derived from Abrams, Feneley, and Torrens (1983).

matical perspective it is more common and usual to model biological oscillators by nonlinear equations. Oscillations in such systems are most often associated with *limit cycle oscillations*, which are oscillations that are usually reestablished following a small perturbation. Nonlinear equations are often needed to represent accurately the complexity and structure of physiological systems.

Recent work has demonstrated that mathematical models for physiological systems that display periodic dynamics can also sometimes display irregular chaotic dynamics in some parameter ranges. As an example, we show in figure 1.8a data from a girl suffering from chronic myelogenous leukemia (CML). CML is a disorder in the production of blood cells (hematopoesis) and generally characterized by a massive increase in circulation of a type of white blood cells called *neutrophils*. In the past two decades, clinical reports have established the existence of an interesting periodic variant in which the peripheral neutrophil counts oscillated around elevated levels with a period of 30–70 days, depending on the patient. A mathematical model for the hematopoetic control system exhibits periodic as well as chaotic dynamics, and an example of chaotic dynamics for this system is shown in figure 1.8b. In chapter 4 we present a summary of mathematical models of biological rhythms which employ nonlinear equations. We show that such models are capable of giving qualitative and sometimes quantitative agreement with observed oscillatory behavior.

We do not attempt a systematic and comprehensive review of all the research dealing with rhythmogenesis because the actual mechanisms for rhythmogenesis in specific systems are generally controversial and not well understood. Furthermore, many of the functionally important characteristics of specific systems—for example, their response to perturbation and mechanisms for initiating and stopping oscillations— can be analyzed in the absence of precise knowledge concerning the mechanisms involved in the generation of the oscillations.

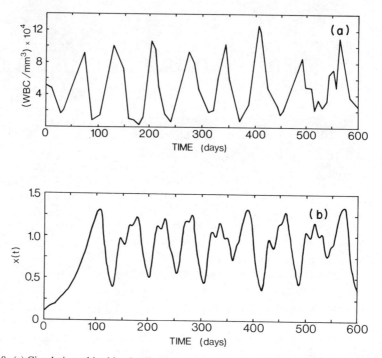

1.8. (a) Circulating white blood cell counts versus time in a 12-year-old girl with diagnosed chronic granulocytic leukemia Redrawn from Gatti et al. (1973). (b) Chaotic dynamics obtained by numerically integrating a nonlinear delay differential equation modeling hematopoetic control. From Mackey and Glass (1977).

1.3 Perturbing Physiological Rhythms

Physiological rhythms do not exist in isolation. Rather, they have multiple interactions among themselves as well as with the external and internal environment. From a functional perspective it is important to analyze the mechanisms involved in initiating and stopping physiological rhythms and the effects of single and periodic perturbations of these rhythms. Clinically, there are many circumstances in which it may be of practical importance to start, stop, or alter a rhythm.

Different physiological rhythms appear and disappear in different stages throughout the lifetime of an individual. Cardiac and respiratory rhythms begin in utero, and even a brief interruption of these rhythms after birth is fatal. The menstrual rhythm is turned on during puberty in females and continues for about thirty-five years, interrupted only by pregnancy and lactation. The strong rhythmic contractions of the

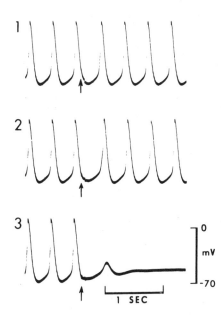

1.9. Phase resetting (lines 1 and 2) and cessation of spontaneous activity (line 3) of SA nodal pacemakers from kittens by brief subthreshold, depolarizing pulses. The spontaneous activity is only annihilated if the stimulus occurs over a narrow range of stimulus amplitude and phase. From Jalife and Antzelevitch (1979).

uterus during labor are initiated naturally about 38–42 weeks after fertilization, or they may be mechanically or chemically induced. Rhythmic peristaltic contraction and daily fluctuations of numerous physiological parameters (e.g., temperature, urine production, blood sugar) take place and we are barely aware of them. Other rhythms such as locomotion, chewing, sleeping, and orgasm usually require active initiation by the individual.

In chapter 5 we discuss several different ways that physiological rhythms can be initiated and stopped. One way by which some rhythms can be terminated is by delivery of a stimulus of critical magnitude delivered at a critical phase of an ongoing rhythm. This is illustrated in figure 1.9, which shows the effect of a depolarizing stimulus delivered to a spontaneously oscillating preparation consisting of sinoatrial (SA) cells (pacemaker tissue) from a rabbit heart. At most phases the stimulus resets the rhythm, whereas over a narrow range of phases the oscillation is annihilated. The recognition that some oscillations can be annihilated by a single stimulus is an important conclusion based on a mathematical analysis of the properties of spontaneously oscillating systems. The experimental observation of the stopping of the cardiac oscillator by a single perturbing stimulus confirmed predictions based on theoretical analysis that such behavior should be possible.

The general problem of analyzing the effects of single as well as periodic stimuli on physiological systems is of key interest for a number of different reasons:

1. In a normal situation the amplitude, frequency, and phase of a biological oscillator are generally under the control of inputs to the

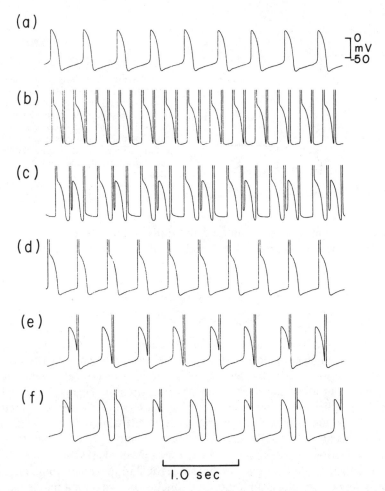

1.10. Recording of transmembrane potential from spontaneously beating aggregates of cells from embryonic chick heart. In lines (b)–(e) the rapid brief upward deflection represents the artifact resulting from periodic stimulation with an intracellular microelectrode. (a) Control; (b) 2:1 phase locking; (c) 3:2 phase locking; (d) 1:1 phase locking; (e) 2:2 phase locking, the stimulus falls at two different phase of the cycle; (f) 2:3 phase locking. From Glass et al. (1984).

oscillator. Thus, characterizing the effects of single and periodic stimuli is of potential functional significance.

2. Pathologically occurring biological rhythms may sometimes be generated or diagnosed by perturbation of an ongoing rhythm.

3. Perturbation of the rhythmic activity of a physiological oscillator can be used to derive information concerning the properties of the underlying oscillation. Conversely, given the mathematical properties of a model oscillator, it is possible to make predictions concerning the expected experimental responses of the oscillator to single and periodic perturbations as the stimulation parameters are varied.

In chapter 6 we consider the effects of single stimuli delivered to a spontaneous oscillation. A single stimulus will generally reset an ongoing rhythm, but the resetting depends on both the amplitude and phase of the delivered stimulus.

In chapter 7 the effects of periodic stimulation of physiological rhythms are considered. In response to a periodic input, a physiological rhythm may become entrained or phase-locked to the periodic stimuli. In this case, there is a periodic rhythm so that for each N cycles of one rhythm there are M cycles of the second rhythm. To illustrate, we show the entrainment of action potentials from spontaneously beating aggregates of ventricular cells from embryonic chick hearts to periodic electrical stimulation (figure 1.10). In addition, at other stimulation frequencies and amplitudes, stable phase-locked dynamics are not observed. Instead, irregular aperiodic dynamics are observed (figure 1.11a). The relative phase of each stimulus is defined to be the time from the preceding upstroke to the stimulus divided by the mean cycle length in the absence of stimulation. If the phase of each stimulus is now plotted as a function of the phase of the preceding stimulus, there is a clear relation between the two (figure 1.11b). In this example the form of the function in figure 1.11b indicates that the irregular dynamics are associated with chaotic dynamics.

1.4 Spatial Oscillations

Throughout most of this book we do not explicitly consider the spatial organization of the oscillations. A usual circumstance is that oscillations spread in an orderly fashion from a pacemaker tissue, as in the heart. However, in some circumstances simple wavelike propagation from a pacemaker is no longer found, and spatial properties of oscillations must be considered. Chapter 8 deals with the organization of oscillations in space.

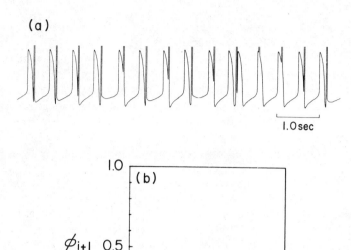

(a)

1.0 sec

(b)

ϕ_{i+1} 0.5

0   0.5   1.0

ϕ_i

1.11. (a) Chaotic dynamics arising in the same preparation as in figure 1.10. (b) A plot showing the phases (determined by measuring the time interval from the upstroke of the action potential to each stimulus and dividing this by the intrinsic cycle length in the absence of stimulation) of successive stimuli as a function of the preceding stimulus derived from the same record from which the trace in (a) was taken. The original observation of chaos in this preparation is due to Guevara, Glass, and Shrier (1981). This figure is from Glass, Shrier, and Bélair (1986).

The key requirement for propagated activity is that the tissue in question be *excitable*. This means that a stimulus (typically a depolarization of neural or muscular tissue) can spread to neighboring tissue in a nondecrementing fashion. Excitability is found in both quiescent tissue and spontaneously oscillating tissue. In both cases, however, there is typically a *refractory time* following excitation during which the tissue cannot be excited. There are consequences of this on both the spatial and temporal properties of excitation. In one dimension this leads to an orderly spread of excitation in the form of a traveling wave. Two waves traveling in opposite directions will not pass through each other, but rather annihilate each other. In addition, as stimulation frequency increases, conduction of each impulse is not possible, and it often happens that some impulses are wholly or partially blocked. Such phenomena are of clinical importance in cardiac tissue and possibly in other excitable tissues such as smooth muscle. In higher dimensions the spread of excitation is potentially more complex, and, in addition to simple waves spreading in an orderly fashion from a point source, other more complex patterns of propagation are observed.

In recent years, progress and interest in propagating waves have been stimulated by studies of the Belousov-Zhabotinsky reaction, a chemical reaction discovered in Russia in the 1950s that displays excitable kinetics. One of the remarkable features of this reaction is that the chemical reactants include a redox indicator, ferroin, which changes color as the reaction progresses. As a consequence, it is possible to see the progress of the reaction. When the reaction is prepared in a petri dish, target patterns consisting of outwardly propagating concentric circular waves are found initially (figure 1.12). These target patterns appear to arise from a pacemaker, which is often a piece of dust or a scratch in the petri dish, and the frequencies of the different pacemakers can vary. In 1972 Winfree discovered that if the dish is tilted, the target patterns may be destroyed and complex one-armed spiral geometries result (figure 1.12). All the spirals in the same dish rotate with the same velocity and can be easily observed visually.

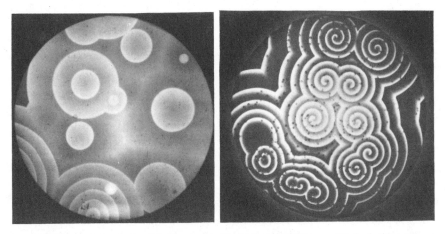

1.12. Waves of chemical activity in an excitable medium, the Belousov-Zhabotinsky reaction. On the left are target patterns that propagate outward from a point source; on the right are spiral waves, which rotate in clockwise and counterclockwise orientations. The first observation of spiral waves of chemical activity was in Winfree (1972). Figures provided by A. T. Winfree.

The importance of this reaction is that it appears to typify the spatial modes of organization in living, excitable biological tissue. For example, it has been suggested that "reentry" phenomena in cardiology are analogous to the spiral morphologies in the Belousov-Zhabotinsky reaction. In three dimensions, even more complex geometric organizations of an excitable medium have been found.

1.5 Dynamical Disease

The human body is a complex, spatially and temporally organized unit. In many different diseases, the normal organization breaks down and is replaced by some abnormal dynamics. We have proposed that these diseases, characterized by abnormal temporal organization, be called *dynamical diseases*. Over the years, this theme has provided a central rationale for our research and for that of others interested in the application of mathematics to the study of human disease.

Throughout the book we offer examples drawn from diverse physiological systems in which abnormal dynamics appears to be associated with disease. In such situations, we anticipate that appropriate mathematical formulations of the relevant physiological control system will display qualitative changes in the dynamics associated with the onset of the disease.

Though it is undoubtedly true that normal, intact physiological control systems can be shown to undergo such transitions in vitro and/or in animal models, we must recognize that in many of the clinical situations in which abnormal dynamics occur there are also preexisting pathological structural abnormalities as well as changes in the physiological control parameters. The interplay between the structural and control parameter alterations in producing abnormal dynamics is a complex problem requiring detailed analysis in individual cases. We believe that detailed mathematical analysis of abnormalities associated with disease is needed, but this view is not currently popular among physicians or medical researchers.

In chapter 9 we deal specifically with the study of dynamical disease and the difficulties inherent in practical application of the concepts of nonlinear dynamics to medicine. From this discussion it will be clear to the reader that the general approach that is taken throughout the book is still at an early stage of development, and that concrete applications to medicine are rudimentary. Yet it is our hope that the potential utility of understanding the dynamics of disease will be evident, and that future researchers will channel their efforts in these directions.

Notes and References, Chapter 1

This chapter provides an introduction and summary of the main themes of the book. For detailed discussions and references, the appropriate chapters of the book should be consulted. There are, however, a number of previous books and reviews which treat specific topics that may be of interest to readers of this book.

The origin of modern nonlinear dynamics is in the original papers of Poincaré (1881, 1882, 1885), which can be found in his collected works (Poincaré 1954). The recent recognition of the importance of chaos was greatly stimulated by the review of May (1976). This paper, as well as many others, can be found in the collections of Cvitanovic (1984) and Hao (1984). There has been a recent proliferation of books on nonlinear dynamics, chaos, and related fields. The texts by Arnold (1983), Guckenheimer and Holmes (1983), Lasota and Mackey (1985), and Devaney (1986) have a rigorous mathematical approach. More physically oriented are texts by Bergé et al. (1984), Schuster (1984), and Thompson and Stewart (1986). An engaging and nontechnical introduction to chaos is given by Gleick (1987). To date there have been no texts dealing with chaos in biological systems. However, earlier treatments of nonlinear oscillations in biology can be found in Pavlidis (1973) and Winfree (1980, 1987a,b). Several recent reviews and collections of papers deal with mathematical approaches to oscillation and chaos in biological and related systems (Olsen and Degn 1985; Holden 1986; Othmer 1986; Koslow, Mandell, and Schlesinger 1987; Rensing, an der Heiden, and Mackey 1987).

The papers of Bernard (1878) and Cannon (1926, 1929), with their discussion and elaboration of the concept of homeostasis, make fascinating reading. A comprehensive listing of papers dealing with physiological rhythms is clearly impossible, but the following references may be consulted as starting points in various areas:

Autonomic nervous system: Polosa 1984
Blood diseases: Wintrobe 1976; Mackey 1979a,b
Cardiac arrhythmias: Bellett 1971; Goldberger and Goldberger 1981
    Marriott and Conover 1983; Zipes and Jalife 1985
Central pattern generators: A. H. Cohen, Rossignol, and Grillner 1988
Circadian: Wever 1979; Moore-Ede and Czeisler 1984; Strogatz 1986
Electrical activity of the cortex: Kiloh et al. 1981
Hormonal system: Knobil 1974; Atwater et al. 1980
Intestinal system: Weisbrodt 1981; Christensen and Wingate 1983; Wingate 1983
Peripheral blood vessels: Siegel et al. 1984
Pupil: Stark 1984
Respiration: Benchetrit, Baconnier, and Demongeot 1987
Tremor: Findley and Capildeo 1984
Ureters: Constantinou 1974

The spatial organization of oscillations, with particular emphasis on chemical and cardiac systems, is extensively reviewed in Winfree (1987b).

An early recognition of the importance of oscillations in human disease can be found in the books by Reimann (1963) and Richter (1965). For the original exposition of the concept of dynamical diseases, see Mackey and Glass (1977), and Glass and Mackey (1979a). Elaborations of this concept are to be found in Mackey and an der Heiden (1982), Guevara et al. (1983) an der Heiden and Mackey (1988), and Mackey and Milton (1987). Several papers dealing with rhythms in psychiatry are collected in Wehr and Goodwin (1983).

Chapter 2

Steady States, Oscillations, and Chaos in Physiological Systems

This chapter provides an introduction to the mathematical concepts of steady states, oscillation, and chaos as applied to physiological systems. In section 2.1 we introduce differential equations and describe the dynamics found in the equation for exponential growth and decay. In section 2.2 we discuss steady states in differential equations, and in section 2.3 we discuss oscillatory solutions of differential equations. The concept that transitions (called bifurcations) can occur between different modes of dynamic organization as parameters are changed is introduced in section 2.4. A different type of mathematical model, called a finite difference equation, can also be used as a model for biological dynamics. In section 2.5 we discuss finite difference equations and illustrate the concepts of steady states, oscillation, chaos, and bifurcation in this class of models. The material in this chapter is of an introductory nature. Readers who are not familiar with this material need not read it in one sitting but can use it as a reference as the need arises.

2.1 Variables, Equations, and Qualitative Analysis

Theoretical analyses of physiological systems attempt to develop equations that describe the time evolution of physiological variables, for example, blood gas concentrations, pupil diameter, membrane potential, or blood cell concentrations. Mathematical models developed to express the time evolution of systems are often written as differential equations, such as

$$\frac{dx}{dt} = f(x), \tag{2.1}$$

where the independent variable x is a function of time t, and dx/dt

represents the rate of change of the variable x (with respect to t) given by the function f. A solution of the differential equation (2.1) gives x as a function of time, designated $x(t)$, starting from some initial condition at $t = 0$, $x(t = 0) = x_0$.

One way to solve a differential equation is by direct integration. As an example, consider the simple differential equation,

$$\frac{dx}{dt} = \lambda - \gamma x, \tag{2.2}$$

where λ and γ are constants (sometimes called parameters). In this equation it is useful to consider λ to be a rate of production and γ to be a rate of decay. From equation (2.2) we find directly that

$$x(t) = x_0 e^{-\gamma t} + \frac{\lambda}{\gamma}(1 - e^{-\gamma t}). \tag{2.3}$$

By substituting equation (2.3) into equation (2.2), an equality is obtained, demonstrating that equation (2.3) is a solution of equation (2.2). The situation with $\lambda = 0$ is the familiar *exponential decay* if $\gamma > 0$, or *exponential growth* if $\gamma < 0$. The graph of the solution of equation (2.2) is shown in figure 2.1 for a situation in which λ and γ are greater than 0. Starting from any initial condition, then $\lim_{t \to \infty} x(t) = \lambda/\gamma$.

In this example, it is possible to integrate the equation directly to give an analytic solution. The physical sciences place great emphasis

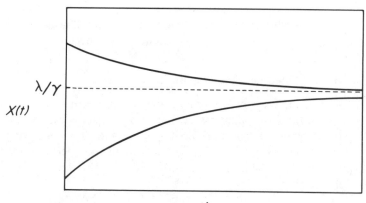

2.1. Exponential decay shown by the solution of equation (2.2) as a function of time starting from two different initial conditions.

on obtaining analytic solutions to differential equations, with a consequent emphasis in applied mathematics courses on mathematical techniques to integrate differential equations analytically. Since biological systems are generally described by *nonlinear differential equations* (i.e., the right-hand side of the differential equation contains nonlinear terms), for which no analytic solution is available, alternative techniques to the analytic integration of differential equations must often be sought in the study of biological problems. Moreover, biological systems are so complex that it is generally impossible to specify exactly the dynamical equations describing the system. Thus the dynamical equations must generally be considered as approximations and may not have the same validity as differential equations in the physical sciences as, for example, Newton's equations, Maxwell's equations, or Schrödinger's equation.

One alternative technique involves the numerical solution of the differential equation, which was pioneered by Hodgkin and Huxley in 1952 in their beautiful studies of the squid axon. Such techniques are now routinely employed to study the properties of equations that model electrical activity of neural and cardiac tissue. In addition to using numerical techniques, it is frequently possible to deduce important qualitative properties of the solutions of nonlinear equations without explicitly solving them. Examples of such qualitative properties are the number and the stability of solutions of the equations. These qualitative analytic methods are of major importance in the analysis of biological dynamics.

2.2 Steady States

The physiological concept of homeostasis refers to the tendency to maintain a relatively constant internal milieu in the face of changing environmental conditions. Homeostasis can be associated with the notion of stable steady states in mathematics. A *steady state* (also called *equilibrium point* or *fixed point*) is a set of values of the variables of a system for which the system does not change as time proceeds. For models formulated in terms of a differential equation like equation (2.1), a steady state x^* is a solution of the differential equation at which $dx/dt = 0$. For example, $x^* = \lambda/\gamma$ is a steady state of equation (2.2). A steady state is *stable* if, after a small perturbation away from the steady state, the solution returns to the steady state as $t \to \infty$. The observation that the blood pressure was reestablished following a small perturbation in figure 1.1 is an indication that the steady state can be described mathematically as a stable steady state. Likewise, for $\gamma > 0$ in equation (2.2), the steady state $x^* = \lambda/\gamma$ is a stable steady state, and following a perturbation away from this steady state, the solution $x(t)$ returns to the steady state as time proceeds. The solution of any one-dimensional differential equation like equation (2.1) will always approach a steady state or become infinite as $t \to \infty$. Starting at some initial point x_0, x will monotonically increase if $f(x_0) > 0$ and will monotonically decrease if $f(x_0) < 0$. The increase (or decrease) will continue until a steady state is reached or until $t \to \infty$.

2.3 Limit Cycles and the Phase Plane

Biological systems do not always approach steady states, but may sometimes oscillate. Many people are familiar with oscillations that can arise in differential equations such as those representing the motion of a pendulum, or a satellite in the earth's gravitational field. In these physical systems, if one ignores friction, then, following a perturbation induced by injecting energy into the system, the oscillation is different from the original oscillation. Thus the amplitude of oscillation of an ideal pendulum (no energy dissipation due to friction) will generally be changed following a perturbation to the pendulum.

However, the situation may be quite different in physiological systems. The effects of perturbation of an oscillating physiological system can be illustrated by considering the effects of a brief electrical shock delivered to an aggregate of spontaneously beating cells derived from the ventricles of an embryonic chick heart. In response to a brief electrical stimulus, there is a resetting of the phase of subsequent action potentials, but the original cycle time is reestablished within several beats, as shown in figure 2.2. The reestablishment of the rhythm following the stimulus indicates that the rhythm is stable. Since the term "steady state" refers to an unchanging state (not an oscillating one) the rhythm shown in figure 2.2 is not a steady state, and another concept is needed.

The requisite concept was provided by Poincaré in his study of differential equations with two variables. In such equations it is possible to have an oscillation that is reestablished following a small perturbation delivered at any phase of the oscillation. Poincaré called such oscillations *stable limit cycles*.

We illustrate the concept of limit cycle oscillations by giving a simple mathematical example. Define a polar coordinate system in which the variable r measures the distance from the origin, and ϕ measures the

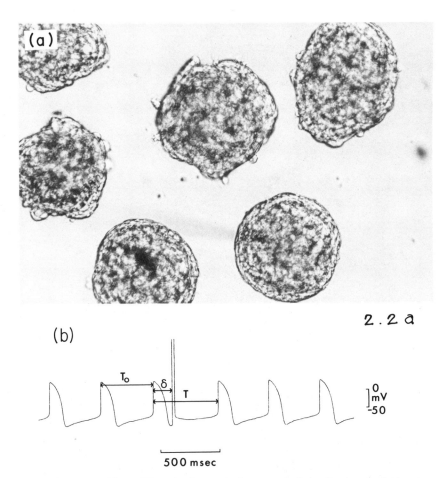

2.2. (a) Aggregates (about 100 mμ in diameter) of spontaneously beating heart cells derived from the ventricles of 7-day-old embryonic chicks. All cells in a single aggregate
are electrically coupled and beat with the same intrinsic frequency. Photograph provided
by A. Shrier. (b) Transmembrane potential from an aggregate showing spontaneous electrical activity and the effect of a 20-msec, 9-nA depolarizing pulse delivered through an
intracellular microelectrode. The control cycle length is T_0 and the perturbed cycle length
T. From Glass et al. (1984).

angular coordinate (see figure 2.3a). Consider this pair of differential
equations:

$$\frac{dr}{dt} = ar(1 - r), \qquad a > 0$$

$$\frac{d\phi}{dt} = 2\pi$$

(2.4)

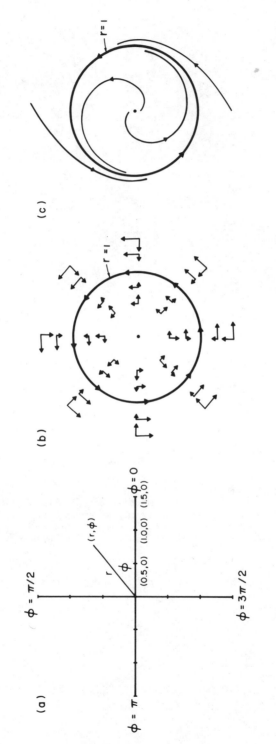

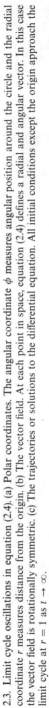

(a)

$\phi = \pi/2$

(r, ϕ)

r

ϕ

$\phi = 0$

$(0.5,0)$ $(1.0,0)$ $(1.5,0)$

$\phi = \pi$

$\phi = 3\pi/2$

(b)

$r = 1$

(c)

$r = 1$

2.3. Limit cycle oscillations in equation (2.4). (a) Polar coordinates. The angular coordinate ϕ measures angular position around the circle and the radial coordinate r measures distance from the origin. (b) The vector field. At each point in space, equation (2.4) defines a radial and angular vector. In this case the vector field is rotationally symmetric. (c) The trajectories or solutions to the differential equation. All initial conditions except the origin approach the limit cycle at $r = 1$ as $t \to \infty$.

In this system of equations, ϕ increases at a rate 2π, and dr/dt is a quadratic function of r, independent of ϕ. For $r(t) > 1$, we have $dr/dt < 0$, and conversely for $0 < r(t) < 1$ it is clear that $dr/dt > 0$. Consequently, for any initial condition $r_0 \neq 0$, $\lim r(t) = 1$. Since this is a polar coordinate system, we can take the values of ϕ modulo 2π, which means that any value of ϕ satisfying $0 < \phi < 2\pi$ is considered equivalent to $\phi + 2n\pi$, where n is a positive integer. Therefore, for any initial condition (except $r_0 = 0$), as time proceeds the solution $r(t)$ will approach the circle defined by $r = 1$, with a period of oscillation equal to 1. The set of initial conditions for which $r(t)$ approaches the limit cycle as $t \to \infty$ is called the *basin of attraction* of the limit cycle.

This dynamic behavior can be graphically viewed in the (r, ϕ) plane by sketching the time evolution of r and ϕ. Equation (2.4) defines the rate of change in the r and ϕ coordinates at each point in the (r, ϕ) space. We represent this rate of change by the vectors determined from equation (2.4) at several points in the (r, ϕ) plane (see figure 2.3b). By taking the resultant vector at each point in space and following from one vector to the next, we can trace out the path followed by r and ϕ as time proceeds. This path, called the *trajectory*, is shown in figure 2.3c, starting from several different initial conditions. The sketch of the dynamics in the two-dimensional plane in figure 2.3c is often called the *phase plane portrait*. From an examination of the phase plane we see that any initial condition, with the exception of the origin $r = 0$, will approach the cycle at $r = 1$ in the limit $t \to \infty$. This cycle is reached in the limit $t \to \infty$ from points not on the cycle at $t = 0$, hence the name *limit cycle*. Limit cycles are not possible in linear systems or in one-dimensional ordinary differential equations.

Since the first description of two-dimensional radially symmetric differential equations with limit cycles was given by Poincaré, we propose that these systems be called *Poincaré oscillators*. If it is assumed that a biological oscillator is associated with a limit cycle oscillation, then it is possible to make a number of predictions concerning the properties of the biological oscillator without knowing the detailed equations of motion (see chapters 5, 6, and 7).

2.4 Local Stability, Bifurcations, and Structural Stability

The local stability of a steady state or limit cycle is determined by perturbing with small stimuli. If a steady state or limit cycle is reestablished, then they are *stable*. If, on the other hand, a small perturbation induces a change in the dynamics so that the original dynamics are not reestablished, then the steady state or limit cycle is called *unstable*. In practical situations, there will always be small environmental perturbations continuously buffeting any biological system, so any observed steady state or oscillation must necessarily be locally stable. In equation (2.4), the origin $r = 0$ is a steady state since at that point $dr/dt = 0$. However, any small perturbation away from this steady state will lead eventually to the stable limit cycle at $r = 1$. Thus the steady state at $r = 0$ is unstable.

In equations describing biological systems there are typically one or more parameters needed to describe the system, the environment, and their interaction. As parameters change, the local stability of the steady states and cycles may change. Any value of a parameter at which the number and/or stability of steady states and cycles change is called a *bifurcation point*, and the system is said to undergo a *bifurcation*. From a mathematical perspective, initiating and stopping oscillations in physiological systems may be associated with bifurcations in the associated mathematical models. As an example, consider the cessation of respiration, which can be brought about by lowering the CO_2 and raising the O_2 by hyperventilation (see chapter 5). In an appropriate mathematical model, there should be loss of stability of an oscillating solution. We say there is a bifurcation in the mathematical model at the appropriate levels of blood gases.

To this point, we have only considered the local stability of steady states and cycles. Another type of stability is associated with the stability of the basic structure of the equations and the biological system they represent. Given an arbitrary small perturbation to a system of equations, if the main qualitative features remain unchanged (i.e., the topology of the system does not change), the equations are called *structurally stable*. Since in biological systems the parameters in the equations for the dynamics of key systems—if they can ever be known—would certainly be different in different individuals, it seems reasonable to assume that these equations for dynamics in physiological systems are structurally stable. The importance of structural stability in mathematical models of biological systems has been forcefully argued by the topologist Thom. Since small changes in parameters lead to qualitatively different dynamics at bifurcation points, systems are not structurally stable at bifurcation points.

2.5 Bifurcation and Chaos in Finite Difference Equations

In figure 1.11 we plotted the phase of a periodic stimulus delivered to spontaneously beating cells as a function of the phase of the preceding stimulus. Thus this system can be approximated by the equation

$$x_{i+1} = S(x_i), \tag{2.5}$$

where x_i is the phase of the ith stimulus, and S is a function that relates the value of x_{i+1} to x_i. Equation (2.5) is a *finite difference equation*.

To illustrate the properties of finite difference equations, we choose a simple functional form for S in equation (2.5). We assume the function S is a quadratic function, so

$$x_{i+1} = ax_i(1 - x_i) \qquad 0 \leqslant a \leqslant 4. \tag{2.6}$$

The graph of this equation is shown in figure 2.4 for several values of a. The single maximum characteristic of this equation is similar to the experimentally determined curve shown in figure 1.11b. Thus it is not unreasonable to expect that an understanding of the dynamics of equation (2.6) might be helpful in understanding the dynamics in the experimental situation shown in figure 1.11. In a finite difference equation, once an initial condition x_0 is chosen, then the subsequent value x_1 can be determined. Then, in a similar fashion $x_2, x_3, \ldots$ can be calculated. The computational process of determining the value x_{i+1} from x_i is called *iteration*. Iteration of finite difference equations may be accomplished either graphically or numerically.

Graphical iteration of finite difference equations is simple and illustrated in figure 2.4. Choose some initial condition x_0, and determine x_1 from the graph. Then this value of x_1 can be used to find x_2 using the same procedure, and the process continues. Graphical iteration is often useful in the first analysis of a problem. If a more exact solution is needed, the finite difference equation can be numerically iterated, and sequences generated by finite difference equations are easily calculated on digital computers. In fact, computer algorithms for the numerical integration of differential equations utilize finite difference equation approximations to the differential equations.

The concepts of steady states and oscillations are also useful in the study of finite difference equations. A *steady state* x^* is a value for which $x_i = x_{i+1} = x^*$ or

$$x^* = S(x^*). \tag{2.7}$$

For the quadratic map of equation (2.6) there are two steady states, $x^* = 0$ and $x^* = (a - 1)/a$. In figure 2.5a the values x approach a steady state $x^* = 1/2$ for $a = 2$ in the quadratic map (equation 2.6). A *cycle of period n* is defined by

$$x_{i+n} = x_i \text{ and } x_{i+j} \neq x_i \text{ for } j = 1, \ldots, n - 1. \tag{2.8}$$

Figures 2.5b and 2.5c illustrate cycles of periods 2 and 4, respectively.

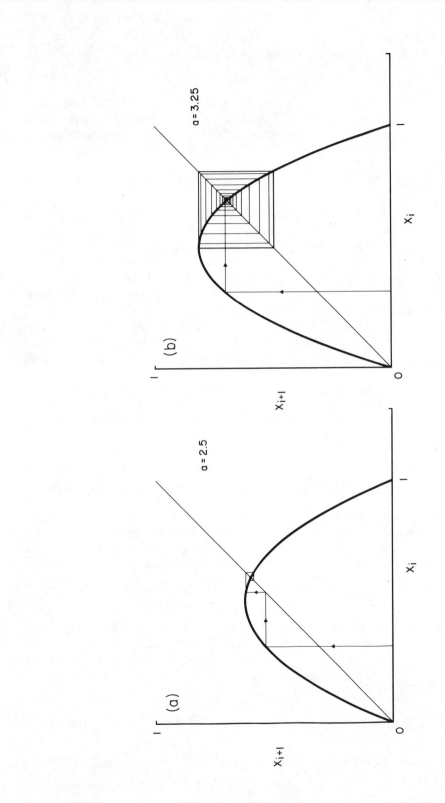

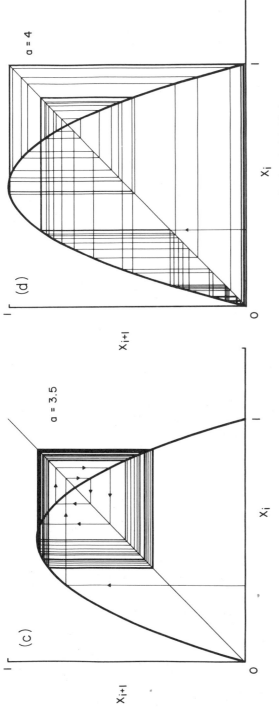

2.4. Graphical solution of equation (2.6). (a) A steady state; (b) a cycle of period 2; (c) a cycle of period 4; (d) chaos.

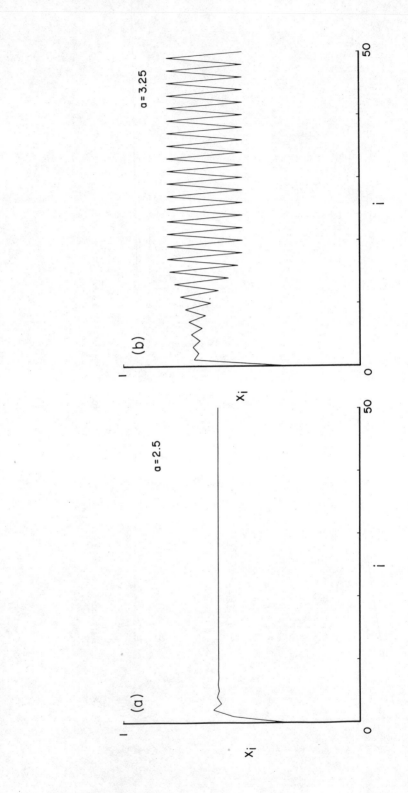

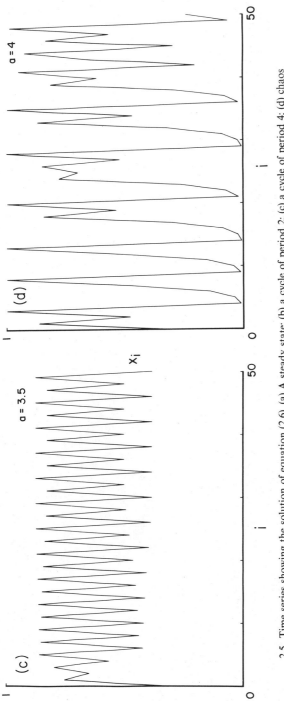

2.5. Time series showing the solution of equation (2.6). (a) A steady state; (b) a cycle of period 2; (c) a cycle of period 4; (d) chaos

Stability of steady states and cycles implies restoration of steady states and cycles, respectively, following a small perturbation.

As parameters change in finite difference equations, *bifurcations* (i.e., changes in qualitative dynamics) are found. One type of bifurcation is a *pitchfork* or *period-doubling bifurcation*, in which a stable cycle of period n becomes unstable and a new stable cycle of period $2n$ is generated as a parameter is varied (see Appendix). One of the remarkable features of the quadratic finite difference equation (2.6) is that as a increases, successive period-doublings occur. Thus, as a increases through the range $3.0 < a < 3.57\ldots$, stable cycles of lengths 1, 2, 4, 8, 16, 32, 64, ... are generated. In figure 2.5 we illustrate cycles of length 1, 2, and 4 (a cycle of length 1 is simply a steady state).

Figure 2.6 shows a *bifurcation diagram* that is generated by numerically iterating equation (2.6) and displaying several hundred successive values of x_i after transients have died away. The range of values of a for which each successive period doubled orbit is found becomes smaller as the length of the period increases, as shown in figure 2.6.

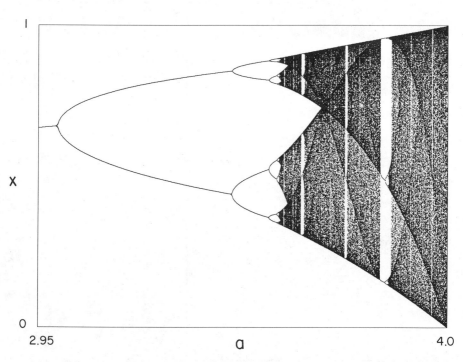

2.6. A bifurcation diagram showing the distribution of the values of x as a function of a for equation (2.6). Figure provided by J. Crutchfield.

Call Δ_{a^n} the range of values of a over which a stable cycle of period n is present. For the quadratic map (equation 2.6) it has been shown numerically that

$$\lim_{n \to \infty} \frac{\Delta_{a^n}}{\Delta_{a^{2n}}} = 4.6692016 \ldots \tag{2.9}$$

It is even more remarkable that this ratio is independent of the precise analytic form of the map, as long as the map has a single quadratic extremal point (maximum). The number $4.6692016 \ldots$ is called *Feigenbaum's constant*.

As a continues to increase in the range $3.57 \ldots < a < 4$, stable periodic orbits with other periods are found (see figure 2.6). These periodic orbits appear in a well-defined sequence called the U (for *universal*) *sequence* (see Appendix). In addition to the stable periodic orbits, "chaotic" dynamics are also observed.

Although a number of mathematical definitions for chaos have been proposed, we try to give the main conceptual notions without technical details. Two main features characterizing chaos must be fulfilled: (1) For some parameter values, almost all initial conditions give rise to aperiodic dynamics; (2) arbitrarily close initial conditions display independent temporal evolution as time proceeds. Thus there is a *sensitive dependence on initial conditions*.

Since an initial condition is known only to some finite degree of accuracy, it is impossible to predict dynamics beyond a certain time in the future since minute differences in initial conditions may have major effects on future temporal evolution. The similarity of the function in figure 1.11b to the quadratic function, combined with the observation of aperiodic dynamics, has led to the interpretation of these dynamics as chaotic (see chapter 7).

Another situation in which many believe that chaotic dynamics can be found is in differential equations modeling atmospheric dynamics. As Lorenz has so delightfully observed, if the solutions of these equations are indeed chaotic, then prediction of the weather too far in advance would be impossible, since an arbitrarily small perturbation (such as the flapping of the wings of a butterfly) could alter the weather on the other side of the globe at some future date. This so-called butterfly effect is often given as a graphic illustration of the operation of chaos. The time scale over which meteorological predictions can be made is not known at this time, though we all have the suspicion that it may be rather short!

In laboratory situations, one can envisage setting up experiments in which measurements must be performed at some later time. If the experimental system was "chaotic," there would be large variability in the measured quantities. It is unlikely that such variability would be considered acceptable by the experimenter (or a reviewer), and initial conditions would likely be manipulated until one obtained "reproducible" results.

2.6 Summary

This chapter provides an introduction to the key mathematical concepts used in this book. Steady states are solutions of differential equations or finite difference equations that are constant in time. Periodic solutions (called cycles or oscillations) of such equations may also be found. Solutions that are reestablished following a perturbation are called stable. In addition, aperiodic, chaotic dynamics, characterized by a sensitive dependence on initial conditions, can be found in both deterministic differential and finite difference equations. Transitions between different modes of dynamic organization are called bifurcations.

Notes and References, Chapter 2

2.1 Variables, Equations, and Qualitative Analysis
Hodgkin and Huxley (1952) called the attention of electrophysiologists to the importance of mathematical modeling and the attention of mathematicians to the richness of the nonlinear problems to be found in biology with their elegant and Nobel Prize-winning study of the process of excitation in the squid giant axon. Their studies included the numerical integration of a partial differential equation using desk calculators. Similar modeling approaches (but using digital computers) have been taken to provide understanding of ionic mechanisms in many other tissues—for example, the process of excitability in cardiac tissue (McAllister, Noble, and Tsien 1975; Noble 1983, 1984) and hormonal release from pancreatic β-cells (Chay and Rinzel 1985). Throughout the text we give many additional examples of nonlinear models in physiology.

2.3 Limit Cycles and the Phase Plane
Our modern understanding of the origin and behavior of limit cycles rests on the seminal work of Poincaré (1881, 1882, 1954) in his study of differential equations with two variables. The systems that we have dubbed "Poincaré oscillators" have also been called λ-ω *systems* (Kopell and Howard 1973). The particular example of a Poincaré oscillator in equation (2.4), called a *radial isochron clock* by Hoppensteadt and Keener (1982), has been used in a modeling context by a number of investigators (Winfree 1975, 1980; Guevara and Glass 1982; Hoppensteadt and Keener 1982; Keener and Glass 1984).

The embroyonic chick-heart preparation, initially studied by DeHaan (1967), has been used as a model system to study cardiac oscillations by a variety of individuals (DeHaan and Fozzard 1975; Scott 1979; Guevara, Glass, and Shrier 1981; Van Meerwijk et al. 1983; Clay, Guevara, and Shrier 1984).

2.4 Local Stability, Bifurcations, and Structural Stability

A good general reference to the concepts in this section is Hirsch and Smale (1974). For an interesting and lively discussion of the potential relevance of the concept of structural stability in biology, see Thom (1970). However, see also Arnold (1983) and the references therein for information about the potential loss of robustness of structural stability as dimensionality is increased.

2.5 Bifurcation and Chaos in Finite Difference Equations

The term "chaos" in its present meaning was introduced by Li and Yorke (1975). However, the significance of such behavior in the natural sciences was recognized earlier; the work of Lorenz (1963) in meteorology is particularly insightful. For references to early studies of the quadratic map, see May (1976). Both Grossman and Thomae (1977) and Feigenbaum (1978) recognized the geometric convergence of the period doubling sequence and numerically computed that the limit in equation (2.9) is 4.6692016 . . . . An analytic estimate of this ratio has been given by May and Oster (1976, 1980). The universal (U) sequence was initially described by Metropolis, Stein, and Stein (1973).

Chapter 3

Noise and Chaos

$\mathbf{E}$xperimental observation of physiological systems often reveals data that are not constant in time and do not show regular periodicities. Such irregularity is often associated with *noise*, or random stochastic fluctuations that are in principle completely unpredictable except in a statistical sense. For example, noise may be due to random thermal fluctuations or other environmental influences. In contrast to noise is chaos, which arises in deterministic systems. In section 3.1 we discuss Poisson and random-walk processes that are normally associated with noise. In section 3.2 we consider chaotic behavior that may be found in simple finite difference equations and show that sometimes such chaotic behavior can mimic behavior normally attributed to noise. We present several examples in which unambiguous interpretation of the dynamics as being either due to noise or to chaotic behavior is not now possible. In section 3.3 we consider several techniques that have been used to identify chaotic dynamics. Two recently applied statistical measures of chaos are the Liapunov number and dimension, which are discussed in section 3.4.

3.1 Poisson Processes and Random Walks

To introduce the notions of noise and random processes, we consider three examples taken from neurophysiology in which dynamics have been modeled by simple random processes.

In figure 1.5 the current flowing through a single membrane ion channel is shown. The discrete jumps are believed to correspond to individual openings of channels in a region of nerve membrane. In figure 3.1a the membrane depolarizations (called *miniature end-plate potentials*) from frog muscle are shown. Each depolarization is associated with the presynaptic release of a quantum containing approximately a thousand molecules of the neurotransmitter acetylcholine. Finally, figure 3.2a shows a record of action potentials in a nerve cell

(a)

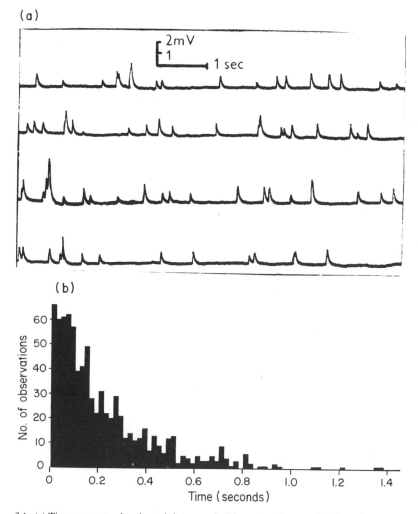

(b)

3.1. (a) Time sequence showing miniature end plate potentials recorded from the neuro-muscular junction of the frog in the presence of 10^{-6} molar prostigmine bromide. (b) Inter-event histogram for miniature end-plate potentials in the neuromuscular junction of the frog. The exponential decay was taken as evidence of a Poisson process. From Fatt and Katz (1952).

in the cochlear nucleus of an anesthetized cat. Simple mathematical models employing random processes have been proposed to explain these data.

The simplest model for a random process is a *Poisson process*. In a Poisson process it is assumed that the probability for an event to occur

(a)

UNIT R-4-10

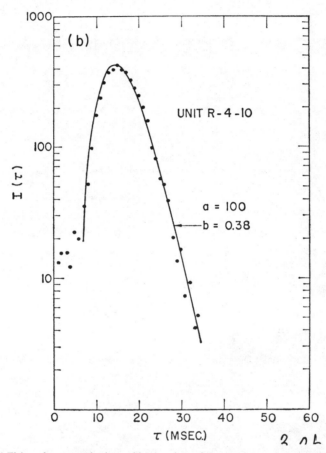

3.2. (a) Firing of a neuron in the cochlear nucleus of a cat under moderate dial-urethane anesthesia. From Rodieck, Kiang, and Gerstein (1962). (b) A fit to the inter-spike histograms for the unit in (a) using equation (3.5). From Gerstein and Mandelbrot (1964).

in a very short time increment dt is $R\,dt$, where the probability R is independent of the previous history, and the probability of two or more events occurring in the increment dt is negligible.

To illustrate a Poisson process, consider a radioactive substance placed in a chamber equipped with a device for detecting and counting

the total number of atomic disintegrations (events) $N(t)$ that have occurred up to a time t. The initial amount of the substance must be
sufficiently large so that there is no significant decrease in the mass
during the time of observation. Clearly, $N(t)$ can only assume integer
values. We repeat this experiment many times, terminating each when
the first disintegration takes place. If we denote the probability that
no disintegrations have taken place in a given experiment up to time
t by $P_0(t)$, $P_0(t) = \mathrm{prob}\{N(t) = 0\}$ and assume that individual disintegrations are independent of each other, then the probability P_0 will
satisfy the differential equation,

$$\frac{dP_0}{dt} = -RP_0(t), \tag{3.1}$$

from which we immediately find (see equation 2.2)

$$P_0(t) = Ce^{-Rt}, \tag{3.2}$$

where C is some arbitrary constant. Since the probability that no disintegrations have taken place at time $t = 0$ is 1, the constant C must
be 1. Thus

$$P_0(t) = e^{-Rt}. \tag{3.3}$$

This type of argument may be continued to derive further properties
of the Poisson process. The probability that there are exactly k events
in a time interval t, $P_k(t)$ is

$$P_k(t) = \frac{(Rt)^k}{k!} e^{-Rt}. \tag{3.4}$$

Moreover, the probability that the interval between an event and the
$(k + 1)st$ following event lies between t and $t + \Delta t$ is $p_k(t)\Delta t$, where

$$p_k(t) = \frac{R(Rt)^k}{k!} e^{-Rt}. \tag{3.5}$$

Equation (3.4) is referred to as the *Poisson distribution*, whereas equation (3.5) is the probability density associated with the Poisson process.
From equation (3.5) it is easy to compute that the average time between events for a Poisson process is $(1/R)$, and the variance is $(1/R^2)$.
Figure 3.3a shows $P_k(t)$ as a function of t for $k = 0$, 1, and 2, while
figure 3.3b shows $P_k(t)$ as a function of k for $Rt = 0.1$, 1.0, and 10.
Experiments in which inter-event histograms are exponentials are
often interpreted to imply that a Poisson process is operating. Thus
channel openings (figure 1.5) and miniature end-plate potentials (figure
3.1) are attributed to the operation of Poisson processes, since the
experimental data are well fit by a single exponential. However, because

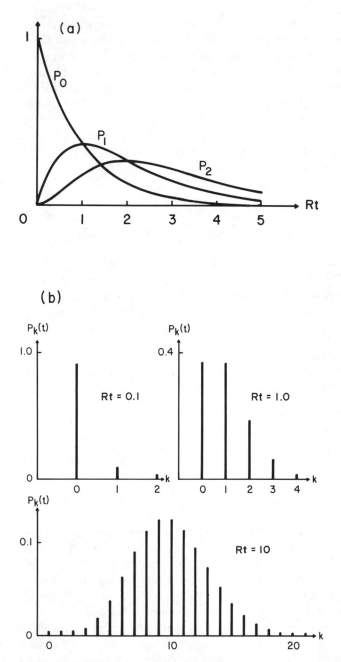

3.3. (a) Probabilities $P_k(t)$ versus Rt for a Poisson process. (b) Plots of $P_k(t)$ versus k for a Poisson process with $Rt = 0.1$, 1.0, and 10. From Lasota and Mackey (1985).

there can be correlations between events that will not be identified from the analysis of the inter-event distribution, it is necessary to determine multiple event statistics but most studies do not undertake such analyses. Although the simple Poisson process is an excellent model for radioactive decay, it is not surprising that statistical analysis of the data in physiological systems often reveals significant discrepancies from the statistics of a Poisson process.

Another model, based on integrate and fire models, can help account for inter-event histograms seen in neuronal activity. In neurons, synapses receive both excitatory and inhibitory inputs from presynaptic neurons. If it is assumed that these inputs arrive randomly and sum linearly over time, and that the changes can be considered to be continuous rather than discrete, then the resulting process is called a *random walk* (or, alternatively, *diffusion, Brownian motion,* or *Wiener process*) (figure 3.4). The distribution of inter-spike intervals corresponds to the distribution of intervals for the first passage of time to the threshold. In this case, the probability density $p(t)$ for the inter-spike intervals is

$$p(t) = kt^{-3/2}e^{[-(a/t)+bt]}, \tag{3.6}$$

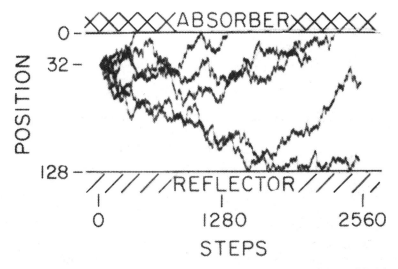

3.4. Typical random walks in one dimension in computer simulation of the model of the firing times for cochlear neurons. From Gerstein and Mandelbrot (1964).

where a is a parameter associated with the difference between threshold and the resting potential, the parameter b reflects differences in the arrival rates of excitatory and inhibitory inputs, and k is a normalization constant. Adjustment of the parameters a and b gives good agreement with the inter-spike interval histogram for a neuron in the cochlear nucleus of the cat (figure 3.2b). Although there is a peak in the histogram at about 15 msec, there is considerable variation in the inter-spike intervals, which is accounted for by the theoretical curve.

3.2 Noise versus Chaos

As we have pointed out, data sets with exponential inter-event histograms are commonly interpreted within the framework of random Poisson processes. However, the distinction between noise and chaos is neither necessarily clear nor sometimes even possible. Indeed, given almost any one-dimensional probability density, it is possible to construct an infinite number of (deterministic) finite difference equations whose iterates are chaotic and which have the given density. This means that observation of a given inter-event density alone cannot be used to infer the dynamics underlying the time series being considered.

As a simple example of this fact, we consider a finite difference equation specifically constructed to have an exponential probability density. Successive iterates of the finite difference equation,

$$t_{i+1} = -\frac{1}{R} \ln\left|1 - 2\exp(-Rt_i)\right|, \tag{3.7}$$

(see figure 3.5a) can be rigorously shown to have the stable density

$$p(t) = Re^{-Rt}, \tag{3.8}$$

which is the same as the Poisson process probability density given in equation (3.5) for $p_0(t)$. Figure 3.5b shows a time series generated from equation (3.7), with $R = 3$. To illustrate that such time series do indeed have the exponential density given by equation (3.7), figure 3.5c shows the numerically determined density of this time series. *Therefore, observation of exponential probability densities is not sufficient to identify a process as a Poisson process.* Plotting a given value as a function of the preceding value may show the underlying structure if the process is generated by a one-dimensional finite difference equation as in equation (3.7). However, identification of chaos in higher-dimensional deterministic systems is difficult, and no simple techniques can be given.

Two biological examples illustrate the problems involved in distinguishing between noise and chaos in data sets having nonexponential

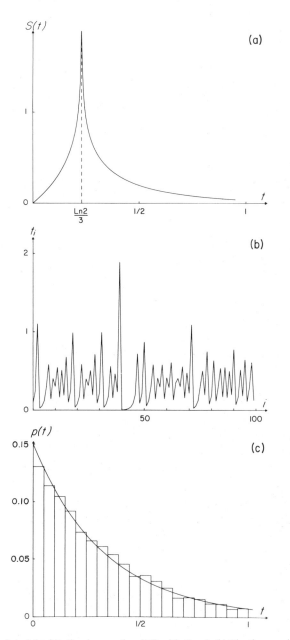

3.5. (a) A plot of the function in equation (3.7) with $R = 3$. (b) The times series generated by iterating the function in (a). (c) The associated probability density. This is a chaotic process with an exponential probability density.

densities. One deals with cell cycle times and the other with the survival of leukemia patients.

From data obtained from cultured cells, it has been proposed that some time after birth a random event takes place within each cell that is necessary for mitosis to take place. In this *random transition cell cycle model*, once the random event occurs, cell division will take place a time T_s later. The probability that a cell drawn at random from a large population has divided at time t after birth, that is, $p_0(t)$, is simply

$$p_0(t) = 0 \qquad \text{for } t < T_s,$$
$$p_0(t) = Re^{-R(t-T_s)} \qquad \text{for } t \geqslant T_s. \tag{3.9}$$

Thus the fraction of cells in the population that has not divided by time t from its birth, denoted by $\alpha(t)$, is

$$\alpha(t) = 1 - \int_{T_s}^t p_0(u)\,du, \tag{3.10}$$

from which we compute

$$\alpha(t) = 1 \qquad \text{for } t < T_s$$
$$\alpha(t) = e^{-R(t-T_s)} \qquad \text{for } t \geqslant T_s, \tag{3.11}$$

which is in fairly good agreement with the data (see figure 3.6a) except in a narrow range of times near T_s. In addition, the data allow one to examine a second statistic, $\beta(t)$, which is the fraction of sister-cell pairs having cell-cycle differing by at least a time t. The random transition model predicts that

$$\beta(t) = e^{-Rt}, \tag{3.12}$$

which is also in agreement with the data (figure 3.6a).

An alternative hypothesis is that there exists a chaotic (but deterministic) intracellular mitotic oscillator that times the cell cycle. Utilizing a simple specific formulation of this model containing only two parameters (remember that the random transition model also contains two parameters, R and T_s), data for the duration of the cell cycle in a variety of cellular populations have been analyzed. In every case, this deterministic chaotic model provides as good a fit to the data, as does the random transition model, shown in the example of figure 3.6b. Is it noise, or is it chaos?

As a second example in which data may be interpreted from either a random or deterministic viewpoint, consider the survival statistics of patients with chronic myelogenous leukemia (CML) following diagnosis

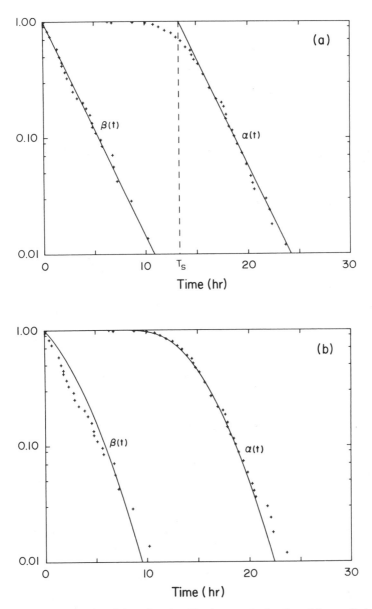

3.6. Statistical properties of the cell cycle. $\alpha(t)$ represents the fraction of those cells in a population that have not divided by time t from their birth as a function of time, and $\beta(t)$ is the fraction of sister-cell pairs having cell-cycle times differing by at least a time t. (a) Fits to the data using the random transition model of Smith and Martin (1973). (b) Fits to the data using the model of Mackey, Santavy, and Selepova (1986). Adapted from Mackey, Santavy, and Selepova (1986).

(figure 3.7). These statistics are customarily described by an exponential function of the form in equation (3.3), or a linear sum of exponentials, even though there is a pronounced hump or shoulder in the data at short times following diagnosis. These exponential fits to the survival data are then taken as an indication that patients die at random with a constant probability per unit time following diagnosis.

As an alternative, a simple finite difference equation model for the production of white blood cells has been proposed. This model incorporates the known feedback control of white blood-cell production with the added assumption that CML is marked by a slow but inexorable increase in the maximal white blood-cell production rate. Under these assumptions, the model indicates that as the disease progresses, an initially stable but slowly increasing white blood-cell density eventually becomes unstable and starts to oscillate wildly, as noted clinically (see figure 1.8). What is even more interesting, once the maximal white blood-cell production rate passes a critical threshold level, the system is predicted to become extinct (the patient dies) at a time that is, in theory, totally predictable given a precisely known initial condition.

However, because of the extreme sensitivity of the evolution of this model to initial conditions, there will be a statistical distribution of

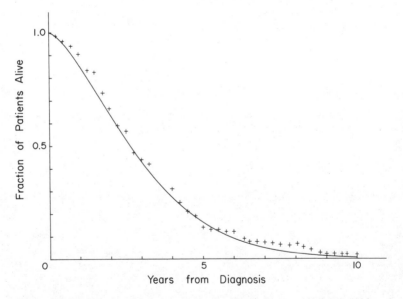

3.7. Survival time statistics for patients with chronic myelogenous leukemia. Although some interpret this to indicate a random death, the data are better fit using the deterministic model of Lasota and Mackey (1980).

the survival times in a population, given a distribution of the patients' clinical states at diagnosis. The model predicts that the fraction of the population of CML patients surviving a time t from diagnosis is equal to $\exp(-kt^n)$, where $n = 3/2$ and k is a parameter depending on the model. Thus the predictions of the model are in very close agreement with the survival data of figure 3.7, which were fit (solid line) by the same function with

$$k = 0.16 \text{ (months)}^{-1.51} \text{ and } n = 1.51.$$

Furthermore, for large populations this surviving fraction is independent of the distribution of initial states.

This prediction has several interesting features. First, it is based on a totally deterministic model for the dynamics of white blood-cell production. Second, it provides a much more accurate description of the available statistical survival data than does the assumption of purely random death at a constant probability per unit time. However, the most interesting aspect is related to the fact that CML survival curves seem to be relatively unaffected by the use of various therapeutic measures involving the use of chemo- and/or radiotherapy. Because these cytotoxic tools will have the effect of resetting the levels of proliferating cell populations, their lack of effect on the survival statistics of a population has a ready interpretation within the context of this deterministic model. That is, the statistical properties of the survival times of a population are insensitive to the distribution of the initial states in spite of the fact that the use of these cytotoxic agents to manipulate these initial conditions may have a dramatic effect on the survival time for a given individual. Specifically, in a given patient the use of these various therapies may dramatically shorten or prolong his or her lifespan.

In conclusion, given some dynamic process—be it deterministic chaos or a stochastic process such as the Poisson process or a random walk—it is frequently possible to compute the associated probability density. However, the inverse operation of determining the dynamic process leading to a given probability density does not have a unique solution. Thus an exponential probability density for inter-event times cannot be used to establish that the underlying dynamic process was a Poisson process without careful assessment of other statistical features of the dynamics.

3.3 Identification of Chaos

The recent widespread recognition that naturally occurring systems (physical, chemical, or biological) can display chaos has led to attempts to identify chaos in the laboratory and in situ. Several hundred papers, mostly after 1980, have been published with this objective. However, the identification of "chaos" in practical situations is difficult. In experimental systems, noise will interact with the dynamics governed by the intrinsic equations determining the evolution of the system. Thus experimental systems by definition have stochastic inputs and present difficulties for theoretical interpretation. The possibility of viewing processes previously classified as noise as deterministically chaotic is confounded by the existence of this noise in the system under study (including measuring devices). Furthermore, even in deterministic equations there are a number of different formal definitions for chaos, and it is important to recognize that chaos is often defined differently from paper to paper. In view of the current difficulties in this area, we briefly describe several different methods now being used to identify chaos.

Power Spectrum

One of the best known and most frequently applied statistical measures to characterize complex time series is the *power spectrum*, which gives a decomposition of a complex time series into a superposition of sinusoidal oscillations of different frequencies. The power spectrum at a given frequency is proportional to the square of the coefficient of the sine wave of that frequency.

Power spectra have been obtained for numerous physiological variables such as heart rate, blood pressure, tidal volume, electroencephalogram, and tremor. Typically, the power spectrum has one or more peaks corresponding to the main frequencies present in the signal. In addition to these main peaks, other frequencies may be present but at lower amplitude, and there is often power over a broad band of frequencies.

Broad-band power spectra, perhaps with superimposed peaks, are often associated with chaotic dynamics. Unfortunately, "noise" is also associated with broad-band spectra, and consequently the presence of a broad-band spectrum is not adequate to establish chaos as opposed to noise.

Poincaré Map

In chapter 2 we discussed the representation of nonlinear dynamics by differential equations. Integration of these equations gives trajectories in phase space. A Poincaré map is established by cutting across the trajectories in a region of phase space with a surface one dimension less than the dimension of the phase space (e.g., with a line if the phase space is two-dimensional). The function that gives the return to this surface on subsequent crossings is a finite difference equation, which is sometimes called the *first return map* or *Poincaré map*.

The Poincaré map derived from a continuous time system can be used to analyze the dynamics. Thus the observation of a Poincaré map consistent with chaotic dynamics is evidence for chaos in the experimental system. The data in figure 1.11 correspond to a Poincaré map for the periodically stimulated cardiac aggregate (see chapter 7). In some systems in which it is difficult or impossible to measure the evolution of all the relevant variables in time, a single variable is sometimes measured, and the value of this variable is plotted as a function of its value at some earlier time. The Poincaré map in this two-dimensional embedding of the time series can then be determined.

Routes to Chaos

We have described the sequence of bifurcations present in the quadratic map as the parameter *a* is varied. In some instances it has been possible to observe the same sequences of bifurcations even in situations in which there is not a well-worked-out theory. For example, observation of a period-doubling bifurcations followed by irregular dynamics is taken as evidence that the irregular dynamics are chaotic.

The strongest evidence for chaotic behavior comes from situations in which there is a theory for the dynamics that shows both periodic and chaotic dynamics as parameters are varied. Corresponding experimental observation of theoretically predicted dynamics, including irregular dynamics for parameter values that give chaos in the deterministic equations, is strong evidence that the experimentally observed dynamics are chaotic. The experiments on periodically stimulated heart cells represent one situation in which such an analysis has been possible (see chapter 7).

Liapunov Number and Dimension

Recent work in nonlinear dynamics has developed quantitative measures for the characterization of complex dynamics. The two most popular of these measures are the *Liapunov number* and *dimension*, which are, respectively, measures of the degree of regularity and the geometry of the dynamics. Although a full discussion of these measures is necessarily highly technical, we briefly discuss them in the next section because of the growing importance of these measures to characterize nonlinear dynamics.

3.4 Strange Attractors, Dimension, and Liapunov Numbers

The mathematical concepts related to the characterization of chaos can be addressed either from the standpoint of differential equations or finite difference equations. In our discussion here, we restrict ourselves to differential equations but note that extension to finite difference equations is possible. An *attractor* is a set of points S such that for almost any point in the neighborhood of S the dynamics approaches S as t approaches infinity. Thus the stable steady states and stable limit cycles discussed in chapters 1 and 2 are attractors. These attractors have a very simple geometric structure (figure 3.8). Specifically, a stable steady state is a point (dimension 0) and a stable limit cycle is a closed nonintersecting curve (dimension 1). It is also possible to have a two-dimensional attractor. An example is an attractor that is a torus (i.e., the surface of a doughnut). The trajectory in this case can wind around the torus an infinite number of times, filling the surface but never intersecting itself (figure 3.8). This situation is called *quasiperiodicity* and is considered in chapter 7.

In the above examples, the attractors have simple geometries with integral dimensions. They are not "strange." Yet it is now widely recognized that attractors can be found which have weird geometrical properties, and in 1971 Ruelle and Takens proposed that such attractors be called *strange attractors*. Since definitions of strange attractors vary, depending on the author, we prefer to avoid a technical discussion of strange attractors but give illustrative examples to give some insight into what the term "weird" geometry means.

D = 0   D = 1

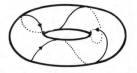

D = 2

3.8. Attractors of integral dimension. For a stable steady state D = 0, for a limit cycle D = 1, and for quasiperiodicity D = 2.

One way to think about strange attractors is to imagine what happens to a little ball of dough embedded in pastry as one makes some delicious *pâte feuilletée* (used in making classic French pastry). To help us visualize the ball of dough, we will dye it purple, but this will not alter its taste in our *gebacken* experiment. The pastry is made by rolling out the dough, covering it with a thin layer of butter, folding it up, and then rolling it out again. The purple ball will get stretched out and twisted around into a very complicated geometry even after a few iterations of the rolling, buttering, and folding process (figure 3.9). However, it is hard to observe this geometry. One way to partially observe the geometry is to slice through the dough to examine the purple layer in some particular cross section. If this were done, there would be some convoluted purple-dough areas in the cross section.

To draw the analogy with nonlinear dynamics, we imagine differential equations with three or more variables. If there are N variables, then any condition is represented by a point in this N-dimensional *phase space* and the evolution in time is represented by a trajectory. One can now take a small volume of points in the phase space (analogous to the purple ball of dough in the example above) and see how the points in the ball evolve and disperse as time proceeds. In the case of chaotic dynamics, the ball of points may eventually stretch out to cover some or all of the attractor and may have a strange geometry. In contrast

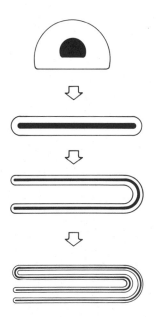

3.9. Schematic picture of the strange geometry encountered in the baking of puff pastry (*pâte feuilletée* As the pastry is rolled out and folded over, a ball of pastry initially in the center gets stretched out into a convoluted sheet.

with the pastry dough, the volume covered by the initial ball of points may decrease (this is what happens in *dissipative* systems).

Recent experimental evidence shows that strange geometries can be observed in experimental systems in physiology. Aihara and colleagues have studied the effects of periodic sinusoidal stimulation of spontaneously oscillating giant neurons from squid. At given phases of the sinusoidal-forcing function, they record the voltage V and its time derivative dV/dt. When this is done for certain stimulation frequencies and amplitudes, they observe complex folded geometries similar to what is observed for strange attractors in nonlinear differential equations exhibiting chaotic dynamics (figure 3.10).

Clearly, it would be nice to have some quantitative method to characterize the geometries of strange attractors. One that has been proposed recently and is now in a state of active development is the dimension. Steady states, limit cycles, and quasiperiodic attractors are associated with an integral dimension (figure 3.8). Yet since early in this century, mathematicians have dealt with pathological sets that are associated with a nonintegral dimension. Such sets have been called *fractals* by Mandelbrot, who has written extensively on the importance of fractals in understanding geometric aspects of natural sciences. In order to

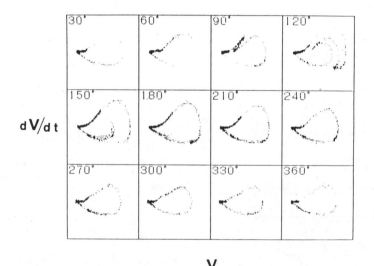

3.10. Stroboscopic plots in which V and dV/dt are plotted at different phases of a sinusoidal forcing of a squid giant axon. The current is 1.5 μA, the forcing frequency is 270 Hz and the intrinsic frequency of the neuron is 200 Hz. From Aihara et al., (1986).

illustrate what the term "fractal" means, we must adopt a definition of dimension, and a number of different ones have been proposed. Probably the simplest is the capacity dimension. Consider a set of points in N-dimensional space. Let $n(\varepsilon)$ be the minimum number of N-dimensional cubes of side ε needed to cover the set. Then the *dimension* of the set is

$$d = \lim_{\varepsilon \to 0} \frac{\log n(\varepsilon)}{\log(1/\varepsilon)}. \tag{3.13}$$

For example, to cover a line of length L, $n(\varepsilon) = L/\varepsilon$ and d is readily computed to be 1. Similarly, for a square of side L we have $n(\varepsilon) = L^2/\varepsilon^2$ and $d = 2$. To visualize a set that has a fractional dimension, consider the construction shown in figure 3.11. Take a line of unit length. Chop away the middle third. Now chop away the middle third of the two remaining pieces. Next chop away the middle third of the remaining four pieces. After this process is repeated an infinite number of times, the set of points remaining is called a *triadic Cantor set*. To compute the dimension, let m be the number of times the cutting operation takes place, so $\varepsilon = (1/3)^m$. If $m = 0$, then $n(\varepsilon) = 1$; if $m = 1$, then $n(\varepsilon) = 2$; if $m = 2$, then $n(\varepsilon) = 4$; and, in general, $n(\varepsilon) = 2^m$. Applying equation (3.13) we readily compute $d = \log 2/\log 3 = 0.630\ldots$.

Nonmathematicians may consider this discussion of fractals to be useless mathematical gibberish. Yet it is now clear that nonlinear systems can have strange attractors that are fractals. Moreover, Mandelbrot and others have proposed that anatomical structures, such as the

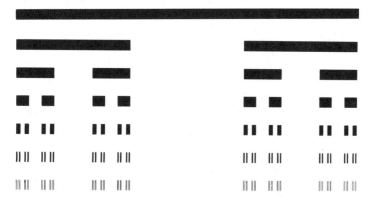

3.11. The triadic Cantor set. Each line is derived from the one above by deleting the middle third of each segment. The dimension is .6309 .... From Mandelbrot (1982).

circulatory system and the lungs, may display fractal geometry. The analysis of fractal aspects of anatomy and dynamics is just beginning and is sure to be an area of much more intensive development.

Another quantitative measure to characterize strange attractors is the Liapunov number, which can be determined from consideration of the evolution of a small ball of points in phase space. As time proceeds the small ball (in N-dimensions) will become ellipsoidal with principal axes $r_i(t)$. Then the Liapunov numbers are

$$\lambda_i = \lim_{t \to \infty} \frac{1}{t} \log_2 \frac{r_i(t)}{r_i(0)}, \tag{3.14}$$

where the $\lambda_i(t)$ are ordered from largest to smallest. Kaplan and Yorke conjectured that the dimension of strange attractors can be computed from the Liapunov numbers.

To this point we have avoided a discussion of the connection between chaos, strange attractors, dimension, and Liapunov numbers. The use of the terms is sometimes confusing and varies from author to author. The approach of Grebogi and colleagues seems clearest. They use the term "chaos" to reflect the dynamics of a system, and "strange attractor" to characterize the geometry of the attractor. A chaotic system is one for which typical orbits on the attractor have a positive Liapunov exponent. A strange attractor has weird geometry such as fractional dimension or nondifferentiability. This distinction seems important because using the above definitions, chaotic dynamics can have attractors that are not strange, and, conversely, nonchaotic dynamics can display strange attractors.

There is currently active research underway on the development of practical algorithms that can be used to compute numerically the dimensions and Liapunov numbers given the values of some variable as a function of time. Applications in biology include analysis of electroencephalograms and electrocardiograms. Unfortunately, the algorithms being used have many potential pitfalls and their convergence properties are currently not well understood. In particular, two aspects still need careful theoretical analysis: (1) the requirements for the size of the data set being analyzed, and (2) the effects of noise, large derivatives, and geometry of the attractor. Because of these difficulties, unambiguous interpretation of published reports is difficult. Any claim for "chaos" based solely on calculation of dimension or the Liapunov number without additional supporting evidence such as well-characterized bifurcations or a believable theory must be viewed with extreme skepticism at present.

3.5 Summary

Random processes are often characterized by their inter-event histograms. For example, it is well known that in the Poisson process the inter-event histogram is an exponential function. We have shown that chaotic systems can also give rise to exponential inter-event histograms. Thus it is not a simple matter to distinguish between noise and chaos, and it is possible that irregular dynamics in many systems that have been ascribed to noise may in fact be due to chaotic processes in deterministic systems. Several methods that have been used to identify chaotic dynamics are discussed. This is an area of very active current research. Clear operational definitions that can be used to evaluate the relative contributions of noise or chaos in a given experimental record are not now available.

Notes and References, Chapter 3

3.1 Poisson Processes and Random Walks

For a discussion of stochastic processes associated with noise, any text in probability theory can be consulted. We recommend Feller (1968). Fitting of interevent histograms to exponentials is a standard procedure in neurophysiology and can be found in many places. We have given examples from Fatt and Katz (1952), who studied miniature end-plate potentials in frog neuromuscular junction, and Sakmann, Noma, and Trautwein. (1983), who studied ion channel openings in heart cells. Van der Kloot, Kita, and Cohen (1975) discuss discrepancies between the predictions of the Poisson process and experimental data for the temporal distribution of miniature end-plate potentials in neuromuscular junction. Gerstein and Mandelbrot (1964) utilized random-walk models for the inter-event distributions of cochlear neurons.

3.2 Noise versus Chaos

The observation that deterministic finite difference equations can give rise to exponential densities is due to Lasota and Mackey (1985). J. A. Smith and Martin (1973) analyzed mitosis using the random transition cell-cycle model. Mackey (1985) and Mackey, Santavy, and Selepova (1986) proposed the alternative hypothesis of a deterministic cell-cycle model. Survival statistics of patients with chronic myelogenous leukemia are taken from Wintrobe (1976) and have been fit by Burch (1976) to an exponential function and a sum of exponentials. Lasota and Mackey (1980) fit these data to a model with deterministic chaos.

3.3 Identification of Chaos

Power spectral analysis is frequently performed in physiology. Representative papers include power spectra of the heart rate (Kitney and Rompelman 1980; Akselrod et al. 1981; Kobayashi and Musha 1982), respiration (Goodman

1964), electroencephalogram (Rapp et al. 1986), and tremor (Findley and Capildeo 1984).

For an early discussion of the separation of noise and chaos in experimental data, see Guckenheimer (1982). A discussion of the different methods being used to determine chaos from experimental data is in Crutchfield et al. (1980), Swinney (1983), and Abraham, Gollub, and Swinney (1984). In addition, recent collections of papers have several contributions dealing with these problems (Cvitanovic 1984; Hao 1984; Mayer-Kress 1986).

For discussions of the use of the Poincaré map to analyze dynamics in differential equations, see Smale (1967), Guckenheimer and Holmes (1983), and Lasota and Mackey (1985). Different routes to chaos are described in Eckmann (1981).

3.4 Strange Attractors, Dimension, and Liapunov Numbers

Since this is a rapidly developing and controversial area, readers will have to look for the most up-to-date information on their own. A start is provided in Ruelle and Takens (1971), Kaplan and Yorke (1979), Farmer, Ott, and Yorke (1983), Grassberger and Procaccia (1983), Grebogi et al. (1984), Eckmann and Ruelle (1985), Wolf et al. (1985), and Kostelich and Swinney (1987). The collection of papers by Mayer-Kress (1986) is an excellent summary of the state of the art in 1986. For computations of the dimension of the EEG, see Babloyantz and Destexhe (1986), Dvorak and Siska (1986), Rapp et al. (1986), as well as several papers in the volumes edited by Mayer-Kress (1986) and Koslow, Mandell, and Schlesinger (1987).

Mandelbrot (1977, 1982) coined the term "fractal," and his enthusiastic celebration of these curious mathematical objects has captured the imagination of mathematicians and physicists. To find out why, consult the beautiful volume by Peitgen and Richter (1986). Attempts to apply these concepts in biology are just getting started (Goldberger et al. 1985; Grebogi et al. 1985; West and Goldberger 1987).

Chapter 4

Mathematical Models
for Biological Oscillations

A large literature exists that proposes many different models for the generation of physiological rhythms. This chapter summarizes the main classes of mechanisms that have been proposed for biological oscillators and illustrates them with representative data. In section 4.1 we describe principles involved in the generation of pacemaker oscillations. Then, in section 4.2, we discuss the generation of motor rhythms by central pattern generators. Two of the proposed mechanisms for central pattern generators are mutual inhibition and sequential disinhibition, which are discussed in sections 4.3 and 4.4. One of the main control mechanisms in the body involves negative feedback. In section 4.5 we show that such systems can lose stability, with resulting stable oscillations. Systems with mixed positive and negative feedback are considered in section 4.6. Such systems can display both oscillatory and chaotic dynamics.

4.1 Pacemaker Oscillations

Many physiological rhythms are generated by a single cell or by electrically coupled isopotential cells that are able to generate oscillating activity in isolation or in the presence of a constant input. We refer to such cells or groups of cells as *pacemakers*.

Pacemaker oscillations are believed to play a role in generating oscillatory behavior in the heart, smooth muscle, many hormonal systems, and neurons. Though we would like to distill the vast literature on pacemaker oscillations into a few simple mechanisms or principles, this is impossible since it appears that there are different mechanisms underlying rhythmogenesis in different systems. It remains unclear whether this reflects the true situation or represents an early assessment that will change in light of future experimental work.

We illustrate the formulation of models of pacemaker oscillations by briefly discussing a mathematical model due to Huxley for the periodic generation of action potentials in the squid giant axon in a low-concentration extracellular calcium solution. As shown by Hodgkin and Huxley in 1952, the action potential is generated as a result of changing time- and voltage-dependent membrane conductances to sodium and potassium. In the squid giant axon, as in other cells, there is relatively high concentration of sodium and low concentration of potassium in the extracellular fluid, relative to the intracellular medium. If the membrane were permeable to only a single ionic species, the transmembrane potential would reflect the net flow of this single species, and the resulting equilibrium potential for that ion could be computed using the Nernst equation. For example, when considering sodium, there would be a net flow of sodium ions into the cell until the electrical potential set up due to the influx of sodium ions counterbalanced the diffusion of sodium into the cell that is due to its concentration gradient. At the Nernst equilibrium potential there is still diffusion across the membrane, but the inward and outward ionic fluxes are equal. By the convention of measuring all potentials relative to the extracellular fluid, the sodium equilibrium potential is positive while the potassium equilibrium potential is negative.

If I is the applied membrane current, then Hodgkin and Huxley assumed that

$$I = C dV/dt + g_{Na}(V - V_{Na}) + g_K(V - V_K) + g_l(V - V_l), \quad (4.1)$$

where C is the membrane capacitance, V is the membrane potential, g_{Na}, g_K, and g_l are the membrane conductances to sodium, potassium, and the leakage ions, respectively, and V_{Na}, V_K, and V_l are the corresponding equilibrium potentials. The richness of the problem arises as a result of the highly nonlinear dependence of the conductances on time and membrane potential.

To characterize these nonlinearities, Hodgkin and Huxley employed the voltage-clamp technique in which a current is applied to maintain the transmembrane potential at a predefined constant value. Under this voltage-clamped condition, $dV/dt = 0$, and it is possible to record single ionic currents by using a variety of pharmacological agents. They showed that during a maintained *depolarization* (positive excursion of the membrane potential away from the resting potential of the squid giant axon membrane, there is a maintained elevation in the potassium conductance and a transient increase in the sodium conductance. Hodgkin and Huxley were able to characterize completely the dependences of the membrane conductances on time and membrane potential, and the brilliance of their work was demonstrated when all of these factors were combined with equation (4.1). With a specified applied current (I), they were able to integrate numerically (using desk calculators!) the resulting set of ordinary differential equations to demonstrate that the equations accurately predicted the time course of the action potential and many other excitable phenomena (figure 4.1).

Thus, under normal conditions, generation of an action potential is accomplished by a depolarization of the membrane, leading to an increase in the membrane conductance to sodium and a consequent increase in the inward movement of sodium. If this sodium influx is sufficient to counteract the effects of the outward movement of potassium ions, it will lead to further depolarization of the membrane past the threshold, and to even more depolarization (the *Hodgkin cycle*). The inactivation of the sodium conductance, in concert with the maintained elevation of the potassium conductance due to the depolarized state of the membrane, results in the return of the membrane potential to its resting value. Under conditions of low extracellular calcium, the relative values of the sodium and potassium conductances are shifted with respect to membrane potential in a way that leads to the spontaneous initiation of the Hodgkin cycle and repetitive generation of action potentials.

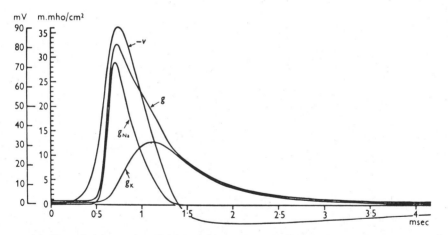

4.1. Numerical integration of the Hodgkin-Huxley equations showing the components of membrane conductance during a propagated action potential. Fom Hodgkin and Huxley (1952).

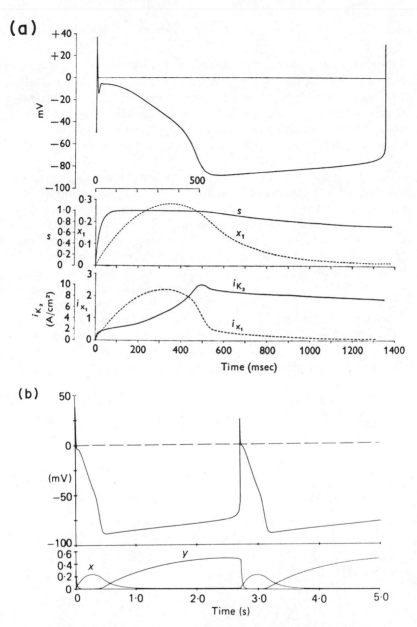

4.2. Calculated action potentials for Purkinje fiber. (a) The action potential in the top panel was initiated by a depolarization to -50 mv. The middle panel shows the time course of Hodgkin–Huxley-type variables x_1 and s, which control outward currents i_x and i_K, respectively (lower panel). The pacemaker depolarization is due to the decrease in the i_K current. Parameter values are appropriate for 2.7 mM $[K]_0$. From McAllister, Noble, and Tsien (1975). (b) Reconstruction of Purkinje fiber pacemaker activity using the DiFrancesco-Noble (1984) equations. Computed action potential is shown above and the computed variations in the gates controlling $i_K(x)$ and $i_f(y)$. Note that the x-variable is comparable to that shown in (a). However, the increase in the y-variable here (controlling inward Na flow) is in complete contrast to the decrease in the s-variable (controlling outward potassium flow) in (a). Parameters values are appropriate for 4 mM $[K]_0$. From Noble (1984).

The techniques developed by Hodgkin and Huxley have been employed in many other systems to study the mechanisms of rhythmogenesis, notably the mechanisms for cardiac rhythmogenesis. In cardiac preparations, the smaller size of cells, combined with the difficulties in obtaining a well-perfused preparation of intact cells, complicate the experiments. There have been major problems associated with defining the mechanisms underlying the depolarizing transition during the so-called pacemaker potentials. In principle, at least two mechanisms can give rise to this transition: (1) a decrease of the outward potassium conductance, or (2) an increase of the inward sodium conductance. In 1975 McAllister, Noble, and Tsien proposed that the main contribution to the cardiac pacemaker potential was associated with the former (figure 4.2a), but current evidence found by DiFrancesco and Noble seems to indicate that the major contribution to the pacemaker potential is due to an inward current carried by sodium or some other ion (figure 4.2b).

The difficulties with the application of the voltage-clamp technique to cardiac cells partially results from the large number of different ionic currents that exist in cardiac pacemakers. Whereas in squid nerve there were two principal currents (sodium and potassium) that could be separated, in cardiac cells there are many more (more than ten), and the experimental and theoretical problems are proportionally greater. Even in nerve, the current experimental evidence indicates the presence of many different ionic mechanisms.

In the past few years, it has become clear, from both experimental and theoretical work, that many pacemakers capable of displaying regular periodic oscillation also may have irregular dynamics, as physiological parameters, or parameters in mathematical models, are varied. To illustrate this interesting behavior, we consider the bursting of pancreatic β-cells. These cells are associated with the secretion of insulin. The bursting behavior of these cells can be monitored with intracellular electrodes. Lebrun and Atwater found that the bursting behavior can be irregular (figure 4.3). Attempts to develop mathematical models of the Hodgkin-Huxley type have been carried out for the pancreatic β-cells. Numerical simulations of these equations by Chay and Rinzel showed that over certain parameter ranges, regular periodic bursting was not found; instead, there were irregular, aperiodic dynamics (figure 4.4). Since these arose in a deterministic mathematical model, it was concluded that this model displayed chaos.

These observations raise important and intriguing problems. The nonlinear differential equations that have been proposed for pacemaker oscillations in diverse tissues may in fact be capable of displaying

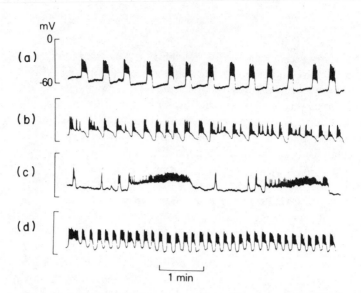

4.3. Bursting activity from four different mouse islets of Langerhans after 30-min exposure to 11.1 mM glucose. (a) Islet from a National Institute of Health mouse. (b) and (c) Islets from Charles River mice. (d) Islet from Charles River mouse after one month on the National Institute of Health mouse diet. From Lebrun and Atwater (1985).

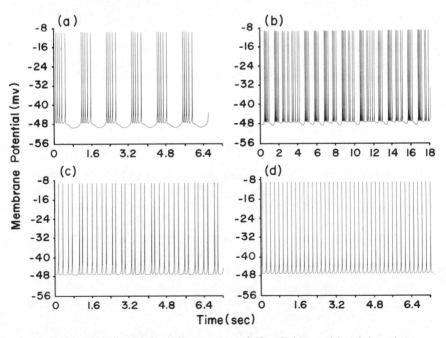

4.4. Various periodic and aperiodic responses of Chay-Keizer model and dependence upon the glucose-dependent uptake rate of intracellular calcium K_{Ca}. (a) Periodic bursting for $K_{Ca} = 0.038$. (b) Chaotic bursting for $K_{Ca} = 0.040$. (c) Chaotic beating for $K_{Ca} = 0.0415$. (d) Periodic beating for $K_{Ca} = 0.045$. Note the altered time scale in (b). Adapted from Chay and Rinzel (1985).

chaotic behavior under changes in parameters associated with differences in the environment of the cells. Indeed, in a collection of papers assembled by Chalazontis and Boisson in 1978, it was shown that pharmacological manipulations of pacemaker cells often resulted in complex rhythms. Thus it appears that complex chaotic rhythms from pacemaker tissue may be more common than is currently recognized.

At the moment, mathematical models for pacemaker oscillations based on realistic ionic mechanisms are so complicated that the only way to determine the model dynamics is by numerical integration. In addition, mathematical models may give excellent reconstructions of action-potential morphology, even in situations in which the ionic mechanisms being modeled have subsequently been found to be in error. Thus, even though it is of major scientific interest to determine the ionic basis of pacemaker oscillation, mathematical modeling of these currents is a difficult task because the resulting models are so complex and there may be alternative interpretations of the same data.

4.2 Central Pattern Generators

There has been great interest since the turn of the century in the mechanisms underlying the generation of motor patterns. It is now clear that motor rhythms in many species are generated in the central nervous system and thus can be maintained even in the absence of sensory input and associated reflexes originating in the periphery. A system generating such a motor rhythm is called a *central pattern generator* (CPG).

Experimental studies of the mechanism of rhythmogenesis by a CPG are generally difficult to perform and interpret. Although it is generally possible to obtain recordings of neural activity which have the same rhythm as the motor output being investigated, it is often not easy to determine if the neural activity represents motor output, sensory input, neural activity from central neurons that are not part of the CPG, or if it really constitutes activity from neurons that are part of the CPG. In addition, synaptic connections between neurons are often difficult to demonstrate experimentally.

In view of the experimental problems involved in defining the CPG, it has proven difficult to develop useful theoretical models. One viewpoint is that, prior to developing theoretical models, it is necessary to have virtually complete information about the neurons in the CPG and their interconnections. Although this may be possible in comparatively simple invertebrate preparations, it is impossible in the vertebrate central nervous system. Furthermore, since most CPGs are composed of many cells, the functional organization of the network is not necessarily transparent once the different cell types and their connections are known. Thus we believe that theory directed at elucidating conceptual aspects of rhythmogenesis is valuable and necessary at the present time.

Two basic classes of models have been proposed for a CPG—the pacemaker and neural network-interaction models. In pacemaker models for the CPG, it is assumed that the rhythm is generated by a cell or small group of cells that oscillate spontaneously. In the case where the pacemaker is a small group of cells, all the cells in the pacemaker are assumed to be synchronized. Although pacemakers have not been found in neural structures generating mammalian motor rhythms, it is very hard to exclude the possibility that such rhythms are generated by pacemaker cells. However, the consensus at the present time is that motor rhythms in mammals are generated by network interactions in neural tissue. The cells in the network may receive tonic input that leads to activity in the cells. The patterning of this activity to generate rhythmic motor output only occurs as a result of network interactions between the different cells or cell groups of the network. In the following three sections we consider several network models that have been proposed for the generation of motor rhythms.

4.3 Mutual Inhibition

Rhythmic motion of a joint is generally accomplished by periodic activation of opposing sets of flexor and extensor muscles. The earliest notions of possible mechanisms generating this rhythmic activity, going back to the turn of the century, assume that there are two groups of neurons. Interactions within each group are excitatory, but the interactions between the two groups are inhibitory. Thus there is a *mutual inhibition*, as diagrammatically represented in figure 4.5. This mechanism was called the *half-center model* by Brown in 1914, but this term is seldom used now.

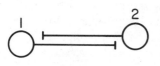

4.5. A model of a neutral network with mutual inhibition. It is assumed that each neuron receives excitatory input and would be tonically active in the absence of the inhibitory synapse. Such a network is not expected to oscillate unless there are some additional factors, such as fatigue or postinhibitory rebound, which will modulate activity in the presence of a constant input.

Assume that there is a tonic input to two mutually inhibitory neurons. Then, depending on the strength of the input and the strength of their inhibitory interactions, one of two possible qualitative dynamics is possible. Both neurons could maintain their tonic activity, but with the activity in each neuron at a lower rate than in the absence of inhibition. Alternatively, one neuron could assume a high, and the other a low, rate of firing. This latter situation is analogous to a phenomenon called *competitive exclusion* in ecology, in which one of two competing species wins out in the struggle for a niche. Such mutually inhibitory interactions can also provide the basis for a neurophysiological "flip-flop" switch.

To generate oscillations from two mutually inhibitory neurons, additional physiological factors such as fatigue, adaptation, or postinhibitory rebound must be included to obtain switching so that first one, then the other neuron is active. As a cell continues to fire at a high rate, metabolites can be depleted, and waste products accumulate corresponding to fatigue of the cell. This fatigue would lead to a diminution in cellular activity. By incorporating fatigue in a theoretical model, it is possible to obtain oscillatory dynamics in a mutually inhibitory network. Alternatively, adaptation would lead to decreasing cellular activity and would also lead to oscillation in mutually inhibitory networks. Another physiological mechanism believed to be important is the excitation often observed following a strong inhibitory input to a neuron. If there is such *postinhibitory rebound*, Perkel and Mulloney showed in 1974 that it is possible to obtain cyclic behavior from mutually inhibitory neurons provided the rebound is sufficiently strong. Such systems can also be found in a stable steady state in which there is low level tonic activity in both neurons. A transition to the oscillating state is accomplished by a strong inhibitory input to one of the cells. The postinhibitory rebound in that cell then leads to a high firing rate, which acts to strongly inhibit the second cell of the network, and a stable oscillation ensues.

A concrete demonstration of oscillations in mutually inhibitory networks is given in studies of the stomatogastric ganglion in lobsters. It is possible to isolate the ganglion from the lobster and to study the cellular and network properties that underlie rhythmogenesis. By injecting photosensitive dye intracellularly and then exposing the cells to intense light, it is possible to selectively kill cells in the stomatogastric ganglion. This elegant technique can be used to isolate two cells, designated PD and LP, which play an important role in rhythmogenesis. If the experimental conditions are correctly manipulated, it is possible to obtain oscillations from the cellular pair even though neither cell alone

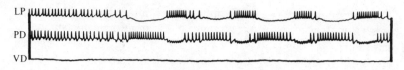

4.6. Alternate bursting observed from the LP and PD neurons from the stomatagastric ganglion of lobsters. These two neurons are mutually inhibitory. This activity is the first clear demonstration of a real "half-center" oscillation in a biological system. From Selverston, Miller, and Wadepuhl (1983).

has a spontaneous oscillation when separated from the other (figure 4.6).

Although mutually inhibitory cell pairs have also been found in other CPGs, the observation of such interactions does not necessarily mean that the mutually inhibitory pair constitutes the CPG. It may simply be one part of the CPG.

4.4 Sequential Disinhibition

In the network with two mutually inhibitory neurons, first one, then the other neuron is active, with the phase switching between them determined by neuronal properties such as fatigue, accommodation, and postinhibitory rebound. However, some neural networks have neurons that are active during the transition times between the two main phases of the cycle. For example, in respiration there are pools of neurons that are active during inspiration and expiration, as well as phase-spanning neurons that are active during the transitions between the two phases (figure 4.7). It is likely that the phase-spanning neurons play a role in the timing of the phases and in the transitions between the phases.

A class of models developed by Kling and Szekeley provide a simple mechanism for rhythmogenesis which extends the earlier mutually inhibitory models and provides for phase switching based on network properties. These models incorporate a mechanism called *sequential disinhibition* or *recurring cyclic inhibition*. An example is shown in figure 4.8. The basic postulate is that there are functionally distinct pools of neurons that receive tonic excitatory input, and that interactions within a single pool of neurons are excitatory. In addition, some of the pools of neurons have inhibitory interactions with neurons in other pools. These networks are capable of generating rhythms in which the activity patterns are determined by the inhibitory interactions between neuronal pools. In these networks, activity in one pool of neurons acts to inhibit the firing of neurons in a second pool. Once the neurons in the second pool of neurons are inhibited, a third pool of neurons, which

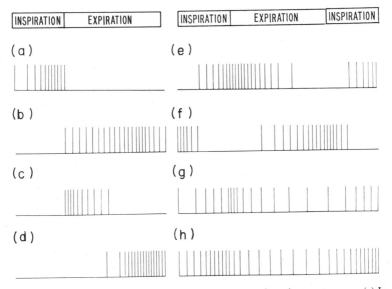

4.7. Schematic classification of major discharge patterns of respiratory neurons. (a) Inspiratory neurons. (b) Expiratory neurons. (c) Early expiratory neurons. (d) Late expiratory neurons. (e) Inspiratory-expiratory neurons. (f) Expiratory-inspiratory neurons. (g) Continuous with peak frequency in expiration. (h) Continuous with peak frequency in inspiration. From M. I. Cohen (1974).

received inhibitory input from the second pool of neurons, can become active. The third pool is thus disinhibited. If the interactions are chosen correctly, stable rhythms will arise. Sequential disinhibition represents a conceptually simple and, to us, an elegant method of generating biological oscillations. Although neurophysiological studies often reveal

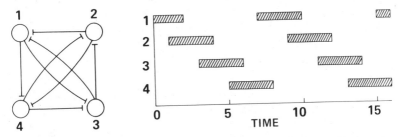

4.8. Schematic representation of a network with sequential disinhibition. Each neuron is assumed to receive tonic excitatory input and would be active in the absence of inhibition. All neuronal interactions are inhibitory. This network tends to be spontaneously oscillatory. The pattern of oscillation shown at the right is similar to the pattern of activity of neurons generating the respiratory rhythm in figure 4.7. From Glass and Young (1979).

that inhibitory interactions do play an important role in rhythmogenesis in neural networks, unambiguous identification of networks displaying sequential disinhibition has not yet occurred.

4.5 Negative Feedback Systems

Those living in temperate climates are familiar with the cycling of heating systems. In the simplest configuration there is a thermostat with a heating element some distance away. The heating element is either on or off, depending upon whether the temperature is less than or greater than some preset value called the set point. After the temperature falls beneath the set point, the heater activates; but because of the time lags in the system, there is some finite period of time before the temperature at the thermostat exceeds the set point. Once this happens, the heating element is turned off, but there will generally be an overshoot of temperature at the thermostat before the heat generated at the heating element is dissipated. It is easy to see how this simple system may cycle, and that the period of oscillation will increase as the distance (equivalent to the time lag) from the heating element to the thermostat increases.

Conceptually simple feedback mechanisms are believed to be fundamental for the control of a large number of different physiological processes. Negative feedback in neural networks may also underlie the organization of CPGs. As a starting point, consider the ordinary differential equation (2.2),

$$\frac{dx}{dt} = \lambda - \gamma x, \qquad (4.2)$$

where x is the controlled variable, and λ and γ are parameters. Since $dx/dt > 0$ for $x < \lambda/\gamma$, and $dx/dt < 0$ for $x > \lambda/\gamma$, this system can be thought of as a simple feedback system with a set point λ/γ. As discussed in chapter 2, equation (4.2) cannot oscillate but will monotonically approach the value λ/γ.

In physiological situations, time lags are often important, and λ and/or γ are not constants but are controlled by feedback mechanisms. To model such systems we assume that λ and/or γ are some appropriate functions of $x(t)$ and/or $x(t - \tau)$ (designated x_τ). Here, τ is a time delay that is used to approximate the time lags present in the physiological system.

Negative feedback systems are those in which deviations in the state variable from a steady-state value tend to be minimized by the feedback. This may result in the approach of the state variable to its steady-state value, or in some circumstances, the emergence of oscillations.

Consider a simple model for the control of ventilation by blood CO_2 levels. Let x denote pCO_2, the partial pressure of CO_2. The CO_2 is eliminated from the body by ventilation and is produced by body tissues at a constant rate λ under constant conditions. The ventilation V is a monotonic increasing function of arterial CO_2 levels some time τ in the past (figure 4.9a). This delay is due to the blood transit time from the brain stem (where ventilation is determined by chemoreceptors and by the "respiratory oscillator") to the lungs (where CO_2 elimination takes place). For computational purposes we assume that

$$V(x) = V_m \frac{x^n}{\theta^n + x^n},\tag{4.3}$$

where V_m is the maximum ventilation and θ and n are parameters used to describe the CO_2 response curve. We also assume that the rate of removal of CO_2 by ventilation is proportional to the pCO_2 multiplied by V. Putting these factors together, we obtain

$$\frac{dx}{dt} = \lambda - \alpha x V_m \frac{x_\tau^n}{\theta^n + x_\tau^n}.\tag{4.4}$$

This is an example of a negative feedback system because increases (decreases) in arterial CO_2 levels lead to increases (decreases) in ventilation, which will, in turn, lead to decreases (increases) in the arterial CO_2 levels.

At the steady state, $dx/dt = 0$. Designate the values of x and V at the steady state by x^* and V^*, respectively and set $S^* = dV/dx|_{x^*}$. Values for these parameters at the steady state, derived from the experimental literature, are

$$\lambda = 6 \text{ mm Hg/min} \qquad x^* = 40 \text{ mm Hg}$$

$$V_m = 80 \text{ liter/min} \qquad V^* = 7 \text{ liter/min} \tag{4.5}$$

$$\tau = 0.25 \text{ min} \qquad S^* = 4 \text{ liter/min mm Hg}.$$

In the Mathematical Appendix we show how the stability of a steady state of a time-delay differential equation may be determined. From this analysis, an approximate criterion shows that the steady state will be stable provided

$$S^* < \pi V^*/2\lambda\tau.\tag{4.6}$$

This analysis also shows that once the steady state becomes unstable, for the parameter values of equation (4.5) there will be an oscillation

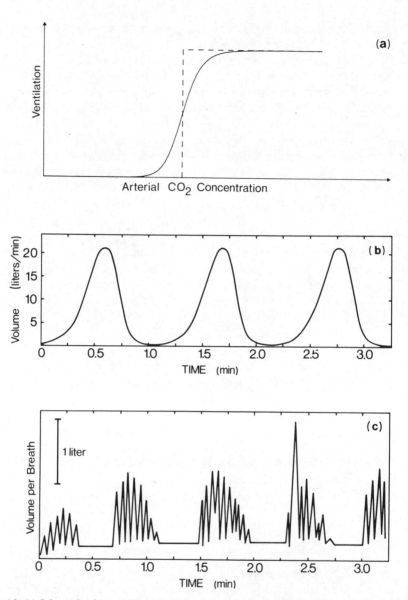

4.9. (a) Schematic picture of the ventilatory control function. (b) Oscillatory behavior of the ventilation obtained by integrating equation (4.4) for parameters in which the ventilation is oscillatory because of instabilities in the negative-feedback control loop. (c) Ventilation during Cheyne-Stokes respiration. Panels (b) and (c) are from Mackey and Glass (1977).

in pCO_2 and, consequently, in the ventilation with a period approximately equal to 4τ (see figure 4.9b).

Equation (4.6) predicts that the steady state pCO_2 level, x^*, may become destabilized in the following ways: (1) if either the slope S^* of the CO_2 control function at the steady state, the time delay τ, or the whole-body CO_2 production rate λ is increased sufficiently, the steady state will become unstable, and (2) if the steady-state ventilation V^* is decreased sufficiently, it may also destabilize the steady state, leading to oscillation with a period of approximately 4τ. This type of transition from a stable steady state to an oscillatory state is called a *Hopf bifurcation* and is discussed in greater detail in chapter 5.

These observations are of interest when considering a breathing pattern known as *Cheyne-Stokes respiration* (see figure 4.9c), in which there is a regular waxing and waning of ventilation. Cheyne-Stokes respiration often occurs in the pathological condition of congestive heart failure (associated with increased circulatory time τ from the lungs to the chemosensitive centers in the brain stem regulating ventilation), in obese individuals (increased τ), and it has been reported after neural brain-stem lesion (associated with increased sensitivity of the ventilatory CO_2 response function, i.e., an elevated S^*). Cheyne-Stokes respiration has been induced in normal dogs via an increase in τ with the addition of an arterial extension, thereby increasing the circulatory time.

In normal individuals, Cheyne-Stokes respiration occurs at high altitude, particularly during sleep. This phenomenon is the cause of the frequently reported inability to sleep soundly during the first few nights following movement to a high altitude from a low altitude. In such circumstances, both O_2 and CO_2 blood gas concentrations are believed to play a role. The low O_2 stimulates hyperventilation, which lowers CO_2 to the lower asymptote of the CO_2 control curve. Ventilation is then sharply reduced or zero until either an increase of CO_2 or a decrease of O_2 stimulates a resumption of ventilation.

Another example of negative feedback is given by studies on the pupil-control system. A small spot of light is shined on the edge of the pupil and stabilized so that it always falls at the same physical location on the eye (figure 4.10a). In response to the spot of light, the pupil contracts. Since light is no longer entering the pupil, the pupil reflexively dilates. Now the light enters the pupil again, leading to pupil contraction. Due to the existing neural-conduction time lags and the nonlinearity of the feedback loop, this system goes into a spontaneous oscillation, which may assume a regular wave form as shown in figure 4.10b.

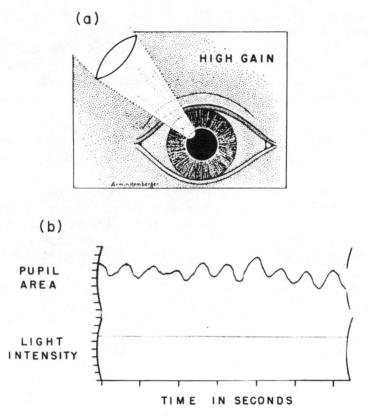

4.10 (a) Illustration of technique used to elicit oscillation in the pupil diameter. Since the light is focused on the border of the iris and the pupil, small movements of the iris lead to large changes in light intensity. (b) Example of spontaneous high-gain oscillations in pupil area obtained with setup in panel (a) with constant light intensity. From Stark (1968).

Other examples in which negative feedback control systems with time delays have been analysed to obtain insight into periodic physiological rhythms arise in hematology, motor control, psychiatry, and the regulation of blood pressure.

4.6 Oscillations in Mixed Feedback Systems with Time Delays

Positive feedback systems are those in which deviations from a steady state tend to be magnified up to a point. Though these tend to be thought of as not playing a significant role in the operation of biological systems because of their presumed unstable behavior, it is common to find situations in which biological feedback incorporates a mixture of positive and negative feedback. This type of feedback is called *mixed feedback*.

As an example of a mixed feedback system in which time delays are important, we consider an apparently simple model for the control of white blood-cell production. It is generally believed that there exists a self-maintaining pluripotential stem-cell population capable of producing committed stem cells for the erythroid (red), myeloid (white), and megakarocytic (platelet) cell lines. With increasing maturation of the myeloid stem cells, they acquire morphological characteristics that allow them to be identified as proliferating myeloid precursors. As maturation proceeds, proliferative activity in these cells ceases, the cells enter a maturational phase, and the nucleus is expelled. Mature neutrophils (one type of white blood cells originating from myeloid stem cells) are released from the marrow into the blood, where they are randomly destroyed with a short half-life (7 hours in humans). The total time required for a recognizable myeloid precursor cell and its progeny to mature and be released is about 6 days in humans.

A hormonal control operates between the circulating neutrophil mass and the myeloid stem cells. Decreases in circulating neutrophil numbers lead to the production and release of the hormone granulopoietin (GP). GP then acts on the myeloid stem-cell population to increase proliferative activity and ultimately increase the flux of cells into the recognizable myeloid precursor compartments.

Let x be the density of circulating neutrophils in cells/kg body weight and γ the random neutrophil destruction rate in hours^{-1}; v the flux of new neutrophils into the blood, measured in cells/kg hour; F (cells/kg hour) the flux of committed myeloid precursor cells into the recognizable myeloid precursor population; and A the cellular amplification that takes place in that compartment (see figure 4.11a). Then, from the preceding description of the neutrophil production system, we have

$$\frac{dx}{dt} = -\gamma x + v(x_\tau), \tag{4.7}$$

where $x_\tau = x(t - \tau)$ and $v(x_\tau) = AF(x_\tau)$ is the current cellular flux into the blood in response to a demand created a time τ in the past.

Over a wide range of circulating neutrophil levels, the neutrophil production rate v is a decreasing function of increasing neutrophil density. However, because of various factors it is expected that at very low neutrophil levels the production rate will fall to close to zero. Thus for

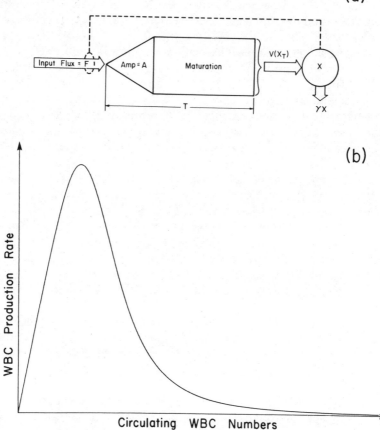

4.11. (a) Schematic diagram for the control of blood cell production. The levels of the circulating blood cells feed back to control the input flux. (b) Schematic diagram of the white blood-cell production rate as a function of the number of circulating white blood cells. This is an example of mixed feedback, and can give rise to chaotic dynamics.

v we pick the humped function (see figure 4.11b),

$$v(x_\tau) = \beta x_\tau \frac{\theta^n}{\theta^n + x_\tau^n}, \tag{4.8}$$

where τ, θ, and n are parameters.

Combining equations (4.7) and (4.8), we obtain

$$\frac{dx}{dt} = -\gamma x + \beta x_\tau \frac{\theta^n}{\theta^n + x_\tau^n}, \tag{4.9}$$

a nonlinear time-delay differential equation describing the dynamics of the circulating neutrophil numbers. Note that in contrast to the ventilatory control system of the previous example, the rate of destruction of cells is now fixed, but the rate of production of cells is under feedback control. Furthermore, instead of having only one steady state, the equation for neutrophil production may have two steady states: $x^* = 0$ and a second steady state,

$$x_0^* = \theta[(\beta - \gamma)/\gamma]^{1/n}, \qquad \text{if } \beta > \gamma. \tag{4.10}$$

As before, the stability of the steady states can be determined. When $\beta < \gamma$ and $x^* = 0$ is the only steady state, it is always stable, as would be expected with the maximum cellular influx less than the neutrophil destruction rate. When $\beta > \gamma$ and two steady states exist, $x^* = 0$ is always unstable while the second steady state given by equation (4.10) may be stable or unstable, depending on the values of the parameters γ, β, n, and τ. The condition for the stability of this second steady state is complicated, and we do not write it down here (see Mathematical Appendix). It suffices to say that increases in β, n, and/or τ may destabilize the steady state, and when this happens the number of circulating neutrophils oscillates with a period between 2τ and 4τ will be the result. Changes in the random destruction rate are more complex, as it turns out that either an increase or a decrease in γ may lead to an instability, depending on the values of the other parameters. This analysis, however, does not even begin to uncover the dynamics that equation (4.9) is capable of producing To explore the dynamics we must abandon our analytic tools in favor of numerical integration.

For normal humans, the following parameter values are estimated from data: $\gamma = 0.1/\text{day}$, $\beta = 0.2/\text{day}$, $n = 10$, and $\tau = 6$ days. With these parameters, the stability analysis predicts that the steady state numbers of neutrophils should be unstable, and indeed numerical integration of equation (4.9) with these values shows a mild oscillation in neutrophil numbers with a period of about 20 days. This period is in agreement with the expected range of the period between 12 and 24 days. In figure 1.8b we carried out the same numerical integration with an increase in τ to 20 days. Note the extreme irregularity of the solution to the totally deterministic equation (4.9). This simple model embodied in equation (4.9) once again illustrates the possibility for the existence of intrinsic chaos in a deterministic, continuous time system. Here the levels of circulating neutrophils in the model are random simply as a consequence of their own evolution equations. As we discussed in chapter 1, the dynamics here shows at least qualitative similarities with data on white blood-cell counts in a patient with chronic myelogenous leukemia. Several investigators believe that CML is generally accompanied by an increase in the transit time τ through the cellular maturation compartments, and thus there may be a connection between the theoretical model and the proposed disease mechanisms.

As yet another example of the generation of periodic and aperiodic behavior from time-delay systems with mixed feedback, consider the process of *recurrent inhibition*, which has been described in almost every type of neural tissue in species ranging from the lowest invertebrates through humans. This process is characterized by presynaptic cells delivering excitation to postsynaptic cells. The postsynaptic cells then generate action potentials, and one effect of these action potentials is to activate inhibitory interneurons via axon collaterals from the postsynaptic cell axons. These interneurons in turn deliver inhibitory activity back to the postsynaptic cells from which their original activation was derived.

A study specifically directed at understanding the dynamics of a recurrent inhibitory circuit in the hippocampus considered the CA3 pyramidal cell, mossy fiber, basket-cell complex (figure 4.12). In this model, formulated within the conceptual framework of equation (4.2), $x(t)$ was identified with the frequency of firing in the CA3 pyramidal cell population. The "production" λ of x is entirely due to the excitatory activity within the mossy fiber population. However, the destruction of x is determined by two different processes: (1) the natural decay of activity that occurs because of the electrotonic properties of the CA3 pyramidal cell membrane, and (2) a "humped" nonlinear type of mixed feedback, of the same type used to describe neutrophil production (equation 4.8), because of the recurrent inhibitory pathway comprised of the basket cells. In addition, there is a time delay in the generation of the recurrent inhibition due to conduction and synaptic delays within the feedback pathway.

The CA3 pyramidal cell–mossy fiber–basket cell complex has been extensively studied, and the relevant parameters for this system are easily estimated. Furthermore, it is known that the inhibitory neural transmitter between the basket cells and the CA3 pyramidal cells is gamma-aminobutyric acid (GABA), and that penicillin binds almost irreversibly to the GABA receptors on the CA3 pyramidal cell mem-

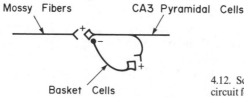

4.12. Schematic figure of a circuit for recurrent inhibition.

brane. Thus, penicillin can be used to titrate the number of available GABA receptors in the postsynaptic cell population, and it is natural to examine the behavior of the model for this system as the GABA receptor density is decreased, corresponding to increasing penicillin levels.

In figure 4.13 we have illustrated the response of this simple model for recurrent inhibition as a function of the number of GABA receptors.

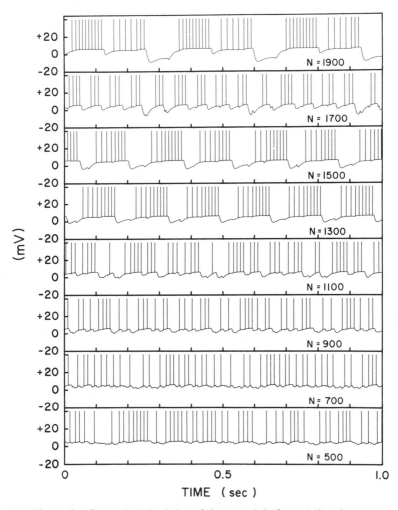

4.13. The results of numerical simulation of the network in figure 4.12. N is a measure of the receptor density. As this density decreases, the rhythm changes from regular bursting behavior with differing periodicities to sustained but irregular firing at low receptor levels. From Mackey and an der Heiden (1984).

As receptor density is decreased to mimic the results of applying penicillin, there is a progressive shift in the cellular activity from regular
burstlike behavior with differing periodicities to a final sustained but
irregular firing pattern at low receptor numbers.

The chaotic behavior of the solutions of time-delay differential
equations with mixed feedback has been seen in models for diverse
phenomena in physiological and ecological systems. Despite the apparent conceptual simplicity and physiological importance of feedback systems incorporating time delays, mathematical analysis of
such systems is extremely difficult. One promising line is to assume
that the nonlinearities in the equations are piecewise linear functions.
In such situations, direct integration of the equations from some initial condition is possible. This technique has been applied to systems
with mixed feedback, and it has been possible to demonstrate a whole
hierarchy of bifurcations between periodic solutions, as well as between
chaotic nonperiodic solutions as parameter values vary. Many other
studies have exploited the use of piecewise constant nonlinearities or
other special types of nonlinearities in time-delay differential equations to obtain insight into the properties of their solutions.

4.7 Summary

Many different approaches to the mathematical modeling of physiological rhythms have been taken. Models for pacemakers attempt to
account quantitatively for ionic currents underlying pacemaker activity.
Since there are a number of different channels, interpretation of experiments and formulation of theoretical models is a complex procedure.
Recent work has demonstrated that slight modifications of the parameters in mathematical models for pacemakers can lead to chaotic dynamics. Mechanisms involving mutual inhibition and sequential inhibition have been proposed for the central generation of motor rhythms
by central pattern generators. Finally, negative feedback and mixed
feedback in systems with time delays can show both oscillatory dynamics and mixed feedback can show chaotic dynamics.

Notes and References, Chapter 4

4.1 Pacemaker Oscillations
The definitive work elucidating the nature of the excitability process in the
protypical excitable tissue, the squid giant axon, was done by Hodgkin and
Huxley (1952). Later, Huxley (1959) extended this work to include the effects

of low-calcium solutions. Many have adapted this approach to characterize other excitable tissues. McAllister, Noble, and Tsien (1975) studied cardiac action potentials, but some of their results have recently been reinterpreted (Noble 1983, 1984; DiFrancesco 1984; DiFrancesco and Noble 1985). Irregular dyanamics in pacemakers have been observed experimentally in a number of neuronal systems (Chalazontis and Boisson 1978), in pancreatic β-cells (Lebrun and Atwater 1985), and following application of drugs to molluscan pacemaker cells (Holden, Winlow, and Haydon 1982). Mathematical models of pacemaker oscillations in neurons (Chay 1984) and pancreatic β-cells (Chay and Rinzel 1985), and in cellular slime molds (Martiel and Goldbeter 1985) also display chaos. There does not appear to be a single common mechanism for pacemaker oscillations (Berridge and Rapp 1979; Noble 1983).

4.2 Central Pattern Generators

The study of central pattern generators (CPGs) goes back at least to early in this century when Brown (1914) demonstrated locomotory movements in cats in which afferent feedback had been abolished. However, a clear theoretical understanding of the mechanisms of CPGs has not yet been obtained, and some have expressed a skeptical opinion on the role of theory (M. I. Cohen 1979). We do not attempt to give a complete review of work on rhythmogenesis in diverse systems but refer the reader to an excellent collection (A. H. Cohen, Rossignol, and Grillner 1988).

4.3 Mutual Inhibition

Conceptual and mathematical models incorporating mutual inhibition have been proposed in diverse contexts (T. G. Brown 1914; Burns and Salmoiraghi 1960; Salmoiraghi and Burns 1960; Harmon 1964; Perkel and Mulloney 1974). Actual demonstrations of oscillations in mutual inhibitory networks have been accomplished by Selverston, Miller, and Wadepuhl (1983) and Satterlie (1985). Mutual inhibition does not necessarily lead to oscillation, but can lead to a situation in which one or the other of the two neurons stays active. A possible role for such a neural switch in memory has been proposed by Hopfield (1984).

4.4 Sequential Disinhibition

The original discussion of sequential disinhibition is in Szekely (1965) and Kling and Szekely (1968). Discussion and analysis of these networks using methods of Boolean analysis is in Glass and Young (1979) and Thomas (1979). Differential equations that model sequential disinhibition and other complex network interactions are in Glass and Pasternack (1978a,b).

For reviews about the mechanisms for respiratory rhythmogenesis, see M. I. Cohen (1974, 1979), Richter and Ballantyne (1983), Euler (1986), and Feldman (1986). These papers indicate that the network-generating respiration is apparently complex, with several different interacting cell types. A proposal that sequential disinhibition networks may underlie respiratory rhythmogenesis is in Petrillo and Glass (1984).

4.5 Negative Feedback Systems

The observation that negative feedback systems may oscillate when the time delays and/or gains are large is well known (Grodins 1963; Milhorn 1966; Stark 1968). However, many treatments of these instabilities rely on methods from linear-systems analysis and thus have limited applicability for analysis of the nonlinear oscillations that can be observed outside of the linear range of the equations. The particular model for Cheyne-Stokes ventilation that we consider is from Mackey and Glass (1977) and Glass and Mackey (1979a), which should be consulted for details on parameter estimation. Other, more complex mathematical models have been developed to account for changes in both O_2 and CO_2, but detailed theoretical analysis of stability properties is generally not possible (Longobardo, Cherniack, and Fishman 1966; Milhorn 1966; Khoo et al. 1982). Clinically, Cheyne-Stokes respiration may be associated with neurological lesions or heart disease (Dowell et al. 1971; Lambertsen 1974). Cheyne-Stokes respiration was induced in dogs by artificially introducing increased circulatory delay following oxygenation of the blood (Guyton, Crowell, and Moore 1956). For a discussion of Cheyne-Stokes respiration at high altitudes, see Waggener et al. (1984). A recent review on breathing control in elderly subjects with extensive references is Pack and Millman (1986).

Oscillatory instabilities in negative feedback systems have been implicated in the genesis of oscillations in many other physiological control systems (Grodins 1963; Milhorn 1966). Other interesting examples have been developed in pupil dynamics (Stark 1968, 1984); periodic autoimmune hemolytic anemia (Mackey 1979b); periodic catatonic schizophrenia (Cronin-Scanlon 1974); the stretch reflex (Lippold 1970); tremor (Merton, Morton, and Rashbass 1967); and in blood pressure (Sagawa, Carrier, and Guyton 1962; Hosomi and Hayashida 1984). Multiple negative feedback with delay can give rise to complex rhythms (Glass, Beuter, and Larocque 1988).

4.6 Oscillations in Mixed Feedback Systems with Time Delays

The model considered here for the control of white blood-cell production is due to Mackey and Glass (1977) and Glass and Mackey (1979a). For a general discussion of the scheme of the construction of the hematopoietic system, see Wintrobe (1976), Mackey (1979a), and Quesenberry and Levitt (1979). Wazewska-Czyzewska (1984) offers evidence that at very low neutrophil levels the production rate will fall to close to zero. See Killman et al. (1963), Ogawa et al. (1970), M. L. Greenberg et al. (1972), and Gavosto (1974) for evidence implicating an increase in the cellular maturation time in chronic myelogenous leukemia.

A number of investigators have evolved models of varying complexity to treat the dynamics of recurrent inhibition (Mates and Horowitz 1976; Kaczmarek and Babloyantz 1977; Traub and Wong 1981; and Knowles et al. 1985). The model considered here is due to Mackey and an der Heiden (1984).

Other investigators have examined the chaotic behavior of the solutions of time-delay differential equations with mixed feedback in models for the control of erythropoiesis (Wazewska-Czyzewska and Lasota 1976; Lasota 1977); in general physiological control systems (an der Heiden 1979); in a model for dopamine dynamics (King, Barchas, and Huberman 1984); in other recurrent inhibition models (an der Heiden, Mackey, and Walther 1981); and in ecology (J. F. Perez, Malta, and Coutinho 1978; May 1980). Use of the technique of replacing nonlinearities with piecewise constant, piecewise linear, or otherwise special functions has been exploited by Glass and Mackey (1979a), Peters (1980), an der Heiden and Mackey (1982), Saupe (1982), an der Heiden (1985), and Walther (1985). This technique has enabled some of these investigators to prove some quite interesting properties of the solutions to these equations. An experimental study which incorporates mixed feedback in the pupil light reflex can be found in Longtin and Milton (1988).

An interesting sidelight is that the equations originally proposed by us to model white blood-cell production have been used extensively by mathematicians and physicists to test algorithms to compute the dimension and Liapunov number (see section 3.4) from a time series (Farmer 1982; Grassberger and Procaccia 1983; Wolf et al. 1985; Le Berre et al. 1987; Kostelich and Swinney 1987).

Chapter 5

Initiation and Termination
of Physiological Rhythms

Transitions between oscillating and nonoscillating states are common in physiological systems. In this chapter we discuss experimentally observed transitions between oscillatory and nonoscillatory dynamics and offer hypotheses to account for these transitions. In section 5.1 we show that an ongoing rhythm may be continuously present but can only lead to readily observable phenomena intermittently. Two distinct routes from oscillatory to nonoscillatory states, known in bifurcation theory as soft and hard excitation, are discussed in sections 5.2 and 5.3, respectively. In the case of hard excitation, there can be two stable dynamic states—one oscillatory and the other not—for a given set of parameters. In this situation, annihilation of the oscillation with a single stimulus should be possible. The observation of such annihilation in experimental systems is discussed in section 5.4.

5.1 Tapping into an Ongoing Oscillation

One possibility for the initiation and termination of biological rhythms is that an underlying rhythm is continuously maintained but that the organism can tap into and out of the rhythm by changing a control parameter. A metaphor is a digital watch that is always running, but whose readout can be turned on and off.

Low-amplitude membrane potential oscillations, sometimes with superimposed action potentials, have been observed in several systems. An important physiological system in which such behavior is commonly found is the gut. Figure 5.1 shows a slow wave oscillation of the smooth muscle tissue. The action potentials present on some of the slow waves are associated with contractions of the smooth muscle (see section 8.1).

The possibility for low-amplitude oscillations, which may be called *slow wave* or *subthreshold oscillations*, may be one of the intrinsic features of pacemaker oscillations, and such oscillations have been observed in many studies. A particularly striking example occurs in

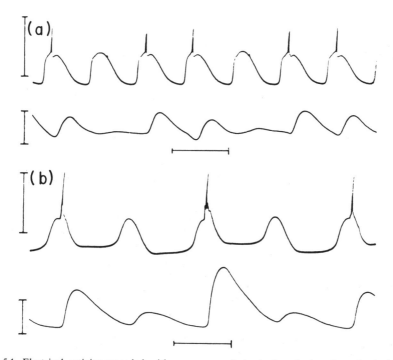

5.1. Electrical activity recorded with a pressure electrode (top line) and mechanical activity (bottom line) recorded from a segment of the cat jejunum. Slow waves that are associated with spike potentials give rise to the largest contractions. (a) Normal activity. (b) Same segment of intestine 40 min after bathing in Tyrode's solution containing 10% of Ca^{2+} present in (a). Note the reduced frequency and apparent increase in magnitude of membrane potential. Time calibration: 5 sec; potential calibration: 0 to -5 mv; tension calibration: 0 to 3 gm, increase in tension upward. From Bortoff (1961).

numerical studies of an ionic model for pacemaker oscillations in Purkinje fiber (figure 5.2). As the magnitude of an injected hyperpolarizing current is increased, a number of different rhythms arise in which low- and high-amplitude oscillations can be observed. In this situation the appearance of an action potential depends on whether or not the oscillating membrane potential exceeds a threshold. If this threshold is exceeded, then the rapid depolarization is observed.

A phenomenon that is at least superficially similar occurs in ovulation. Some patients who exhibit the *luteinized unruptured follicle syndrome* have apparently normal menstrual cycling (or ongoing oscillation), but the key event of rupturing the follicle and release of an ovum (or ovulation) does not always take place. These examples show that in some systems an underlying oscillation can continue even in the absence of what many would assume is the key or critical event.

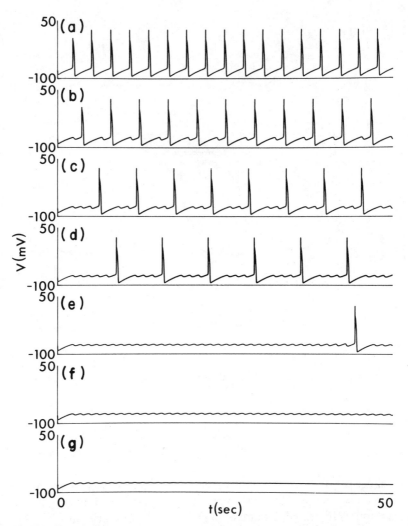

5.2. The effect of injecting a constant hyperpolarizing current in the McAllister, Noble, and Tsien (1975) model for Purkinje fiber. The hyperpolarizing current ($\mu A/cm^2$) is (a) 3.3, (b) 3.34, (c) 3.36, (d) 3.37, (e) 3.3785, (f) 3.3787, (g) 3.4. There are normal action potentials in (a), combinations of sub- and superthreshold oscillations in (b)–(e), sub-threshold oscillations in (f), and quiescence in (g). From Guevara (1987).

5.2 Soft Excitation

A natural way for a mathematician to think about turning oscillations on and off is via a parameter-dependent dynamical system. At some parameter values the oscillation is present, whereas at others the oscillation is absent. The simplest situation is schematically depicted in figure 5.3. For parameter values $c < 0$, there is a single steady state, which is stable and globally attracting. However, for $c > 0$ the steady state has become unstable, and there is now a stable limit cycle oscillation. A concrete example of a pair of differential equations displaying this bifurcation is

$$\frac{dr}{dt} = r(c - r^2)$$

$$\frac{d\phi}{dt} = 2\pi. \tag{5.1}$$

Using the qualitative analysis sketched out in chapter 2, we can easily see that for $c < 0$ there is a single steady state, and for $c > 0$ there is an unstable steady state at $r = 0$ and a stable limit cycle with $r = c^{1/2}$. As c is increased through the value $c = 0$, a limit cycle appears, initially at small amplitude and finite frequency (figure 5.3). Although this bifurcation was known to Poincaré, it is common to call it a *supercritical Hopf bifurcation* or *soft excitation*.

As we have already discussed in chapter 4, instabilities in negative and mixed feedback systems may arise when the time delays and/or

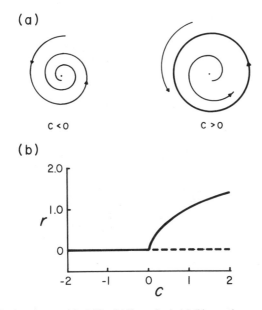

5.3. Soft excitation (supercritical Hopf bifurcation). (a) Phase plane representation of equation (5.1). As c increases, a stable limit cycle oscillation of low amplitude and finite frequency arises. (b) The bifurcation diagram for soft excitation. The solid curve represents the stable solutions and the dashed curve the unstable solutions.

gains increase beyond certain critical values. At the points of the instabilities, supercritical Hopf bifurcations are observed. Such instabilities
are believed to underlie the oscillations observed in Cheyne-Stokes respiration, some types of muscle tremor, many oscillatory hematological
disorders, and the pupillary oscillatory reflex elicited by a spot of light
on the edge of the pupil. In these various situations, an important class
of experiments motivated by the theory would be to determine carefully
the amplitude and frequency of the oscillation as a function of stimulus
intensity near the transition to the stable oscillation. These data could
then be related to the experimental observations by using appropriate
theoretical tools. Unfortunately, such a protocol has not been undertaken frequently in the study of physiological oscillations. Consequently,
many of the examples we give are of only a tentative nature because
more systematic experimental studies are needed.

An interesting and unusual example that may display a supercritical
Hopf bifurcation is provided by studies on the contractility of the uterus
in dysmenorrhea (figure 5.4). It is known that menstruation is associated
with laborlike contractions, but there is disagreement about the particular patterns associated with pain. In a class of patients with pain-
ful menstruation and large-amplitude contractions, the intrauterine
pressure was measured with a catheter at the peak of the discomfort.
A nonsteroidal anti-inflammatory drug (piroxicam) was administered
orally and intrauterine pressure was monitored for the next hour. For
one patient, as shown in figure 5.4, the large-amplitude periodic (or
perhaps "chaotic") contractions initially present subsided, and this was
associated with a diminution of the discomfort. Relief was obtained in

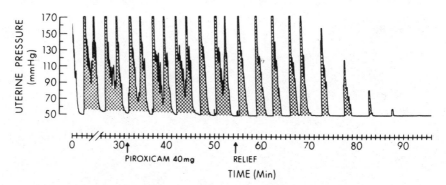

5.4. Uterine pressure as a function of time in a woman with dysmenorrhea. Following
administration of a nonsteroidal, anti-inflammatory drug (piroxicam), the pressure waves
decreased. From Schulman, Duvivier, and Blattner (1983).

69% of the women screened. The decrease of amplitude illustrated in figure 5.4 is precisely what one expects in the supercritical Hopf bifurcation as the control parameter passes its critical value (for example, as c decreases in figure 5.3). However, a word of caution is in order, as dysmennorhea is a complex and poorly understood phenomenon that can probably be caused by many factors. Also, large-amplitude uterine contractions are not unequivocally associated with uterine pain. Thus, although it is possible that the transition observed can be associated with a Hopf bifurcation, the clinical significance is still open to debate.

Soft excitation occurs in systems other than negative and mixed feedback systems, for example, sequential disinhibition. Because sequential disinhibition has been proposed as a basis for rhythmogenesis in the respiratory system and locomotion, it is of interest to examine experimental results on the initiation and termination of the respiratory and locomotory rhythms to see if they display soft excitation as parameters are changed.

A detailed understanding of the control of the initiation and termination of ventilation is not presently available. However, in the context of hyperventilation, the soft excitation scenario seems the most likely. Figure 5.5 shows the phrenic-nerve (the motor nerve for the diaphragm) activity in a mechanically ventilated, anesthetized, paralyzed cat following hyperventilation. The gradual buildup of activity in the phrenic nerve appears consistent with soft excitation. Local cooling of the ventral surface of the medulla or hypoxia produces transitions from respiratory activity to inactivity that are much more abrupt. Figure 5.6 shows a tracing of phrenic activity in an anesthetized, paralyzed, bilaterally vagotomized, and sinoaortic denervated cat in which the common carotid arteries were ligated. At the first arrow, both vertebral

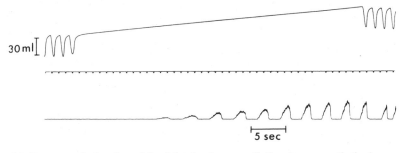

5.5. Recovery of phrenic activity following hyperventilation in an anesthetized, paralyzed cat. Top tracing is lung volume and the bottom tracing is the integrated phrenic neurogram. The offset in lung volume while the ventilator is turned off is artifact. Tracing provided by G. A. Petrillo. From Glass (1987).

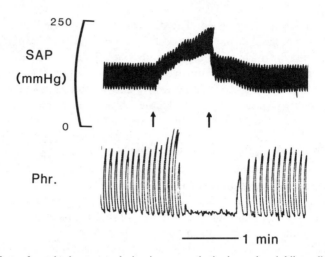

5.6. Effects of vertebral artery occlusion in an anesthetized, paralyzed, bilaterally vago-tomized cat in which the common carotid arteries were ligated. The occlusion was maintained for the time period between the arrows. Top tracing is systemic arterial pressure (SAP), and bottom tracing is the integrated phrenic neurogram. Tracing provided by M. Bachoo and C. Polosa. From Glass (1987).

arteries were occluded, shutting off blood flow to the respiratory centers in the brain stem. This was followed by an immediate increase in blood pressure and a slight increase in phrenic amplitude, followed by an abrupt period of apnea (cessation of breathing). At the second arrow (about 1 min later), normal blood flow was allowed to resume in the vertebral arteries by release of the occlusion clamp, and this led to a rapid onset of periodic phrenic-nerve activity. Our colleagues at McGill, C. Polosa and M. Bachoo, have suggested to us that the abrupt cessation of respiration in this experimental situation may be a model of clinically observed respiratory arrest secondary to cardiac arrest.

As another example we discuss studies on the control of locomotion. Generally, rapid initiation and termination of locomotion is needed in most nonsessile animals in order to survive. In experimental preparations, it is not clear how to reproduce the volitional aspects associated with locomotion. However, in studies in the lamprey it has been found that either electrical stimulation or direct application of pharmacologically active compounds can be effective stimuli in inducing locomotion (or fictive locomotion). As an example, consider the induction of fictive locomotion in the lamprey by electrical stimulation (1 ms pulses at 20 Hz) of the brain stem (figure 5.7). A 3 μA current was not sufficient

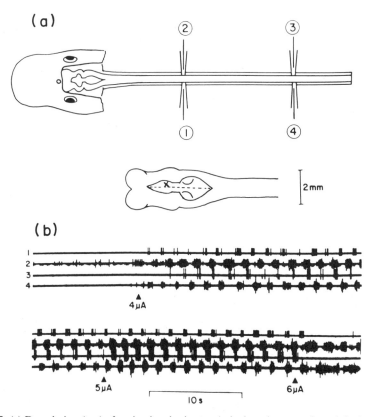

5.7. (a) Dorsal view (top) of an in vitro brain-stem/spinal-cord preparation of the lamprey. "Fictive" swimming behavior was recorded from ventral roots with suction electrodes (1–4). Enlargement of the brain stem (bottom), showing the region of stimulation (x, about 100 μm deep) that produced the records in (b). (b) Continuous record showing the ventral-root motor activity elicited by the stimulation. Increasing the current intensity activates the swimming motor activity. From McClennan and Grillner (1984).

to induce locomotion, but a 4 μA did induce rhythmic activity in the ventral motor roots. At this "threshold" current, bursting began immediately or within a few seconds, and the first burst could occur on either side. The burst activity was controlled by the magnitude of the stimulus.

However, this information on the control of respiration and locomotion is inadequate to characterize the bifurcations underlying the transition from quiescence to rhythmic activity. It is necessary to probe carefully the dynamics as control parameters are both increased and decreased. Are the thresholds for rhythmic activity the same as stimulation intensity is either increased or decreased, or are they different? In soft excitation, there is no hysteresis, and the measured threshold will be the same both for increasing and decreasing control parameters. In the alternative situation, in which there is hysteresis, another mechanism is possible.

5.3 Hard Excitation

In the soft excitation scenario, there is a stable limit cycle that arises initially with low amplitude as a parameter increases. In an alternative scenario, as a parameter increases there suddenly arises a stable, large-amplitude oscillation. One way in which this can occur is called the *subcritical Hopf bifurcation* or *hard excitation*. We give a concrete mathematical example and then describe several physiological examples.

Consider the pair of differential equations,

$$\frac{dr}{dt} = r(c + 2r^2 - r^4)$$

$$\frac{d\phi}{dt} = 2\pi. \tag{5.2}$$

A sketch of the phase plane for this system is shown in figure 5.8a. By setting $dr/dt = 0$, it is readily shown that the solutions for equation (5.2) are given by $r = 0$ and $r^2 = 1 \pm (1 + c)^{1/2}$. The real nonnegative solutions are shown in figure 5.8b. Using the qualitative analysis outlined in chapter 2, we find that the branch $r = 0$ is stable for $c < 0$ and unstable for $c > 0$. For $c > -1$, $r = [1 + (1 + c)^{1/2}]^{1/2}$ is a stable limit cycle. For $-1 < c < 0$, $r = [1 - (1 + c)^{1/2}]^{1/2}$ is an unstable limit cycle.

Now consider what happens as c is increased from negative values. Initially there is a single stable steady state at $r = 0$ (no oscillations). When $c > -1$, there is also a stable limit cycle, but the dynamics will be stuck at the stable steady state. However, as c continues to increase to $c > 0$, the steady state at $r = 0$ becomes unstable, and there is a sudden jump to the stable limit cycle. Thus, as the parameter value passes its critical value, a large-amplitude oscillation appears. If c is now decreased, the large amplitude oscillation will persist until $c < -1$ and will then disappear suddenly (without the amplitude approaching zero). Thus, for $-1 < c < 0$, there is the possibility of two different stable dynamics. The one that is observed will depend on the history of the stimulation (*hysteresis effect*). This simple example makes it clear why careful experimental studies performed with both increasing and decreasing control parameters are needed in order to sort out qualitative properties of the onset and termination of oscillations. The hard

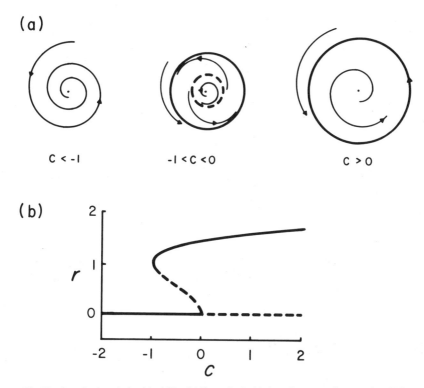

5.8. Hard excitation (subcritical Hopf bifurcation). (a) As c increases in equation (5.2), a stable limit cycle oscillation of finite amplitude and frequency suddenly appears. For intermediate values of c there are two stable behaviors for fixed c, a stable steady state and a stable limit cycle. (b) The bifurcation diagram for hard excitation. The solid curve represents the stable solutions and the dashed curves the unstable solutions.

excitation scenario provides a possible explanation for the threshold shown in figure 5.7.

Probably the most completely analyzed biological system in which hard excitation has been demonstrated is the squid giant axon. Extensive numerical simulations of the Hodgkin-Huxley equations demonstrated that it should be possible to select values for the applied current for which there coexist a locally stable steady state and a locally stable limit cycle, as in figure 5.8. An experimental test of the theoretical prediction, carried out on a space-clamped squid giant axon in low-calcium solution, confirms this prediction.

Two separate types of experiments were carried out. In one, an increasing and decreasing ramp-current stimulus was applied to the nerve. The observation of hysteresis in the firing of the nerve (figure 5.9a) was expected, based on an analysis of the dynamics of this system

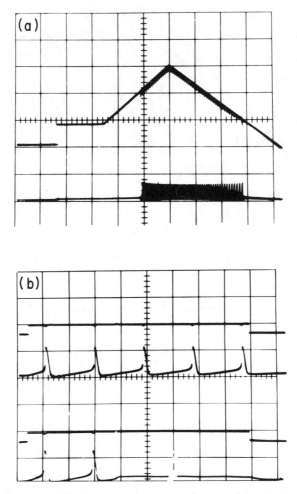

5.9. Activation and annihilation of periodic activity in space-clamped squid axons bathed in a low-calcium artificial seawater solution. (a) A continuous increase and decrease in the current delivered to the axon (top trace) leads to firing of the axon (bottom trace), but the current intensity for the onset and offset of the firing are different, showing a hysteresis effect. Calibration—current: 1 μA/div; membrane voltage: 100 mv/div; time: 50 msec/div. (b) Annihilation of repetitive firing in the axon. Stimulation of the axon with a barely suprathreshold step of current leads to repetitive firing in the top half of the figure. In the bottom half of the figure, repetitive firing is once again induced by a current step, but this is now annihilated by a brief depolarizing perturbation. Calibration—current: 2 μA/div; membrane voltage: 100 mv/div; time: 2 msec/div. From Guttman, Lewis, and Rinzel (1980).

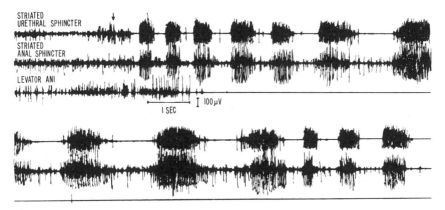

5.10. Electromyographic record from muscles of the pelvic floor of a healthy male volunteer during ejaculation induced using a method attributed to Onan. The arrow indicates a signal from the subject when first noticing a sensation of orgasm. From Petersen and Stener (1970).

that showed the presence of a hard excitation. A remarkable theoretical prediction is that a brief stimulus of critical size delivered at a critical phase of the cycle should be capable of annihilating the oscillation for the situation where a stable steady state coexists with a limit cycle. Such a stimulus was observed in the second type of experiment (figure 5.9b). The annihilation of oscillations in such circumstances is an important discovery and is discussed in more detail in section 5.4.

We close with an additional example which, based on rather incomplete experimental studies, might be associated with hard excitation, but much more detailed analyses are clearly required. Recorded electromyographic activity from pelvic-floor muscles, activated during male orgasm induced by masturbation, shows that at the onset of orgasm a low-level activity in the striated urethral sphincter and striated anal sphincter gives rise to distinct burstlike activity (figure 5.10). The bursts are of irregular duration, and it is reasonable to question whether a limit cycle or some other dynamics would be most appropriate to associate with the bursting activity. However, the sudden onset with no obvious periodicity before orgasm is consistent with hard excitation.

5.4 Annihilation of Limit Cycles: The Black Hole

One characteristic of the hard excitation scenario is that parameter values exist for which there is bistability. Either a stable limit cycle or a stable steady state can be present, depending on the stimulation history (figure 5.8). In this case the steady state is locally stable, and there is a finite surrounding region that is attracting for this stable steady state. Such regions have been called a *black hole* by Winfree. If a perturbation is delivered to the limit cycle, it would be possible to knock the stable limit cycle out of its basin of attraction and into the black hole, resulting in the annihilation of the oscillation.

A stable oscillation of membrane electrical activity is found in experimental studies of the squid giant axon in low extracellular calcium (section 4.1). In 1979 Best carried out detailed quantitative studies of the phase-resetting behavior in the Hodgkin-Huxley equations (appropriately modified to model this experimental system) at different stimulus strengths and showed that for narrow ranges of stimulus amplitude and phase it is possible to perturb the oscillation into its "black hole," thereby annihilating the oscillation. As discussed above, subsequent experimental demonstration of this phenomenon by Guttman and colleagues in 1980 (figure 5.9b) provided a striking theoretical confirmation of the theory. Both depolarizing and hyperpolarizing pulses, which would lead to pacemaker annihilation if delivered at critical phases of the cycle, were found. There is a similar finding in SA node pacemaker cells (see figure 1.9).

Clearly, once an oscillator has been perturbed into its "black hole," it should be possible to reinitiate oscillations if a sufficiently large stimulus were delivered. Curiously, experimental evidence for this transition has not been given.

There are many possible implications of pacemaker annihilation in the interpretation of normal and pathophysiology. At present they are tentative and highly speculative. The following are offered as a sampling.

1. In sinus arrest the sinus pacemaker of the heart ceases to function. This may arise as a consequence of phase resetting of the sinus pacemaker, for example by a vagal burst. In normal individuals, the sinus pacemaker does not appear to be susceptible to such an annihilation, but it is possible that in pathological conditions it would be.

2. Castellanos and coworkers have found electrocardiographic records in patients with ventricular ectopic beats which suggest that a sinus beat at a critical phase of the ectopic cycle can annihilate an ectopic pacemaker. However, it is not clear why subsequent sinus beats do not act as stimuli to reexcite the ectopic pacemaker.

3. Triggered automaticity (initiation of a chain of action potentials by a depolarizing stimulus) has been observed by Wit and Cranefield in cardiac muscle. This phenomenon might be associated with pertur-

bation of the cardiac muscle out of its black hole into a stable periodic orbit.

4. One hypothesized mechanism for sudden infant death syndrome without apparent organic pathology is the cessation of respiration due to a perturbation that annihilates the respiratory rhythm.

5. In amennorhea, the menstrual rhythm is absent but hormonal and surgical treatments can sometimes be given which will restart the rhythm and induce ovulation. The converse situation is also possible. In the absence of knowledge concerning the topological structure of the phase space of the menstrual rhythm, it is conceivable that a black hole exists. If this is the case, then it has been suggested that birth control could be accomplished via perturbation of the menstrual cycle into its black hole.

Though the above are clearly speculative, physiological rhythms that are essential for life must be stably maintained even under the onslaughts of numerous fluctuations and perturbations in the physiological environment. If the sinus pacemaker had a large, easily accessible "black hole" in its phase space, we might not live very long. Thus one might expect that in the course of evolution, survival required that essential oscillators be stable with a large basin of attraction. However, it is also important to recognize that if biological oscillators can really be associated with limit cycle oscillations, then there is the potential for annihilating the oscillation if parameters assume values for which there are black holes.

5.5 Summary

In this chapter we describe four different methods for turning oscillations on and off.

1. The oscillations are always present but at a subthreshold level. In some circumstances the oscillations can become superthreshold, leading to periodic observable events or observable events that occur intermittently but in which the time interval between subsequent events is a multiple of the period of the underlying oscillation.

2. *Soft excitation.* As a parameter increases, a quiescent state becomes unstable, and oscillations, initially with low amplitude, build up. Decrease of the parameter leads to a loss of oscillation following a reversed path with no hysteresis.

3. *Hard excitation.* As a parameter increases an initially quiescent state becomes unstable and large-amplitude oscillations are observed. Decrease of the parameter leads to an abrupt loss of oscillation, but

there exists a range of parameters in which either oscillation or quiescence occurs, depending on the history of the stimulation.

4. In the hard-excitation scenario for parameter ranges with two stable behaviors—one oscillatory and the other not—transitions between the two can be accomplished by a single pulse. However, annihilation of the oscillation by a single pulse can be accomplished only if the pulse is of correct magnitude and delivered over a narrow range of oscillator phases that leads to annihilation and not phase resetting.

Notes and References, Chapter 5

5.1 Tapping into an Ongoing Oscillation

Subthreshold oscillations in neural preparations have been observed by many, including Arvanatiki (1939), Brink, Bronk, and Larrabee (1946), Huxley (1959), Guttman and Barnhill (1970), Guttman, Lewis, and Rinzel (1980), and Holden, Winlow, and Haydon (1982). Rapp and Berridge (1977) have suggested that some pacemaker oscillations (in neural and other tissue) may be due to action potentials riding on top of these subthreshold oscillations. However, Huxley (1959) took the position that they are two entirely different modes of oscillation. Related phenomena in cardiac preparations have been described by Jalife and Antzelevich (1980), Guevara, Shrier and Glass (1986), and Guevara (1987). The luteinized unruptured follicle syndrome is described in Marik and Hulka (1978) and Daly et al. (1985). In a completely different context, Kauffman and Wille (1975) suggested that an underlying mitotic rhythm in *Physarum* can continue even in the absence of mitosis, but this interpretation has been disputed by Tyson and Sachsenmaier (1978) and Loidl and Sachsenmaier (1982).

5.2 Soft Excitation

For a discussion of soft and hard excitation, see Marsden and McCracken (1976), Plant (1982), Arnold (1983), Guckenheimer and Holmes (1983), and Glass (1987). Supercritical Hopf bifurcations occur in negative feedback systems (Langford 1977) and mixed feedback systems (Kazarinoff and Van den Driessche 1979).

The relationship between contractions of the uterus and menstrual pain is a controversial topic that has been much debated (Bickers 1941; Reynolds 1965; Schulman, Duvivier, and Blattner 1983). Analysis of the initiation and termination of contractions of the uterus both under normal and pathological conditions would undoubtedly prove to be of great interest.

Glass and Pasternack (1978b) have studied a theoretical model for sequential disinhibition, and Petrillo and Glass (1984) proposed it as a model for respiratory rhythmogenesis. Transitions of respiratory activity have been produced by Cherniack et al. (1979), who investigated the effects of local cooling of the ventral surface of the medulla; Rohlicek and Polosa (1983) and Millhorn et al. (1984), who studied the effects of hypoxia; and Yamashiro et al. (1985) who studied the effects of hyperventilation. Szekely (1965) and Kling and Szekely (1968) have proposed an important role for sequential disinhibition in the control of locomotion. Grillner (1981) offers a good review of the control of locomotion. Transitions of locomotory activity in the lamprey have been induced by either electrical stimulation or direct application of pharmacologically active compounds (A. H. Cohen and Wallen 1980; Grillner and Wallen 1984; McClennan and Grillner 1984).

5.3 Hard Excitation

Numerical predictions of the coexistence of a locally stable steady state and a locally stable limit cycle in the Hodgkin-Huxley equations by Cooley, Dodge, and H. Cohen (1965), Best (1979), and Rinzel and Miller (1980) were confirmed experimentally by Guttman, Lewis, and Rinzel (1980). Many other fascinating phenomena associated with an abrupt onset of rhythmic activity can be found, but the mechanisms have not yet been enlightened by theoretical analysis. As an example, postmenopausal women sometimes have periodic hot flashes that may have an abrupt onset and offset (Kronenberg et al. 1984). Two other rhythms that may show hard excitation are mastication (Dellow and J. P. Lund 1971; J. P. Lund and Dellow 1973) and masturbation (Petersen and Stener 1970).

5.4 Annihilation of Limit Cycles: The Black Hole

Important insights into the possibility for annihilation of oscillations by a perturbation comes from work by Winfree (1973b, 1977). Following Winfree's suggestion that annihilation of oscillations may be possible by a single well-timed stimulus, such a behavior was observed by Jalife and Antzelevich (1979, 1980) and Guttman, Lewis, and Rinzel (1980) in experimental systems. This work was followed by a clinical study in which electrocardiographic evidence suggested annihilation of an ectopic pacemaker by a sinus beat (Castellanos et al. 1984). A related phenomenon is triggered automaticity, in which a single beat stimulus initiates periodicity (Wit and Cranefield 1976). Glass and Winfree (1984), Paydarfar, Eldridge, and Kiley (1986), and Paydarfar and Eldridge (1987) have hypothesized a link between sudden infant death syndrome and a "black hole" perturbation that annihilates the respiratory rhythm. Tulandi (1985) discussed the pharmacological and surgical restarting of the menstrual rhythm and induction of ovulation in anovulatory humans. Black-hole birth control was first suggested by Winfree (1973b).

Chapter 6

Single Pulse Perturbation
of Biological Oscillators

In this chapter we describe experimental and theoretical results concerning the observed effects of single stimuli delivered to biological oscillators. In section 6.1 we summarize the results of such studies and, in particular, describe experiments on the perturbation of the respiratory and cardiac rhythms to illustrate the main points. In section 6.2 we consider the treatment of phase resetting experiments using integrate and fire models, and in section 6.3 we describe phase resetting in limit cycle models. Several potential applications of phase resetting techniques in diverse fields are discussed in section 6.4. The theory for phase resetting in limit cycle oscillations utilizes a topological approach. The practical problems associated with application of this approach are discussed in section 6.5.

6.1 Overview of Experimental Results

In this section we briefly discuss phase resetting of the respiratory and cardiac rhythms, and then summarize the main properties that have been found experimentally in other systems.

As an illustration of the type of data obtained from perturbation experiments, consider the perturbation of the respiratory cycle by a lung inflation. The respiratory cycle is subdivided into the inspiratory phase, when phrenic activity is strong and the lungs are inflating, and the expiratory phase, when phrenic activity is weak or absent and the lungs are deflating. The inspiratory duration (T_I) and the expiratory duration (T_E) are both modified by lung inflation. Last century, Hering and Breuer demonstrated that lung inflation delivered during inspiration serves to shorten the inspiratory phase of the respiration cycle, whereas lung inflation delivered during expiration prolongs expiration. These Hering-Breuer reflexes are normally operative and are mediated by vagal afferents. If the vagus nerve is sectioned, the frequency of respiration decreases and the tidal volume increases. In current research, the activity of the central respiratory-rhythm generator is often monitored by recording inspiratory-promoting activity from the phrenic nerve, which innervates the diaphragm.

Systematic studies of the Hering-Breuer reflexes have been made by studying the effects of lung inflation at different phases of the respiratory cycle in anesthetized, spontaneously breathing cats while recording phrenic nerve activity. If a lung inflation is delivered during inspiration, and provided the lung inflation is sufficiently large, then inspiration will be terminated. If the volume threshold ($V_{T,eq}$) necessary to terminate inspiration prematurely is plotted as a function of $1/T_I$, the results are as shown in figure 6.1a, and it is observed that subthreshold pulses have little effect on the inspiratory duration. Complementary experiments, measuring the duration of the perturbed expiratory time relative to control as a function of time during expiration at which the stimulus was delivered (figure 6.1b), show that expiratory duration, and hence the perturbed cycle duration, is prolonged by lung inflations delivered during expiration. Although the studies illustrated in figure 6.1 do not address the long-range phase shifts induced by the perturbations, it is known that the phase is reset by stimulation of pulmonary afferent fibers. This means that following several cycles, the firing of phrenic bursts is altered from what it would have been in the absence of stimulation.

As a second example, consider perturbation studies of excised canine Purkinje fibers, which are part of the cardiac conduction system. Once removed from the heart, the Purkinje fiber is threaded through three chambers. In chamber 1 spontaneous oscillatory activity was induced by perfusing with low-potassium Tyrode's solution containing epinephrine. Chamber 2 contained a dextrose and calcium chloride solution that blocked active wave propagation but allowed electrotonic spread of activity. Chamber 3 contained a high-potassium Tyrode's solution to increase the period of the spontaneous oscillation and still allow wave propagation. This "sucrose gap" preparation was taken as an in vitro model of an ectopic, that is, abnormal, pacemaker site (chamber 1) capable of responding to electrical stimuli (chamber 3) via electrotonic interactions transmitted through a region of depressed conductivity (chamber 2).

Electrical stimuli were delivered to chamber 3 at various intervals throughout the pacemaker cycle of chamber 1. When the electrical stimulus was delivered between 0 and 230 msec following an action potential, there was little change in the cycle length. When the stimulus was delivered 800 msec following an action potential, the cycle length was increased 23% over control, and when the stimulus was delivered

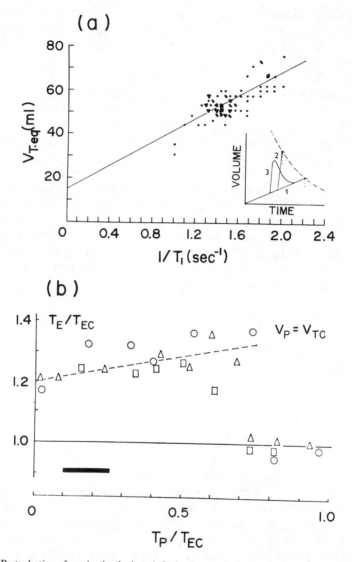

6.1. Perturbation of respiration by lung inflation in anesthetized, spontaneously breathing cats. (a) Brief inflation pulses superimposed upon spontaneous breaths during inspiration. The lung volume at the end of inspiration as measured from the phrenic neurogram is $V_{T,eq}$ and the duration of inspiration is T_I. An inflation pulse either caused inspiration to terminate abruptly (trace 2 in the insert) or, if below the volume threshold, had no appreciable effect (trace 3). From Clark and Euler (1972). (b) Brief inflation pulses delivered during expiration in three cats. The expiratory duration, T_E, divided by the control expiratory duration, T_{EC}, is plotted as a function of the time elapsed since the start of expiration, T_P, divided by T_{EC}. The pulse was of control tidal volume size and its duration is approximated by the solid bar. From Knox (1973).

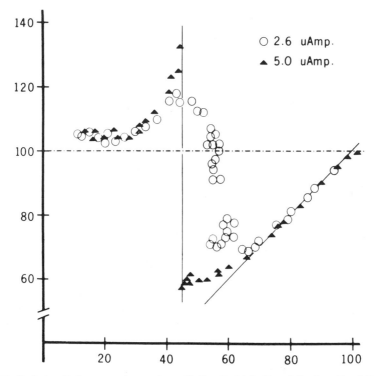

6.2. Cycle length from spontaneously oscillating Purkinje fiber following stimulation with brief electrical current pulses, as described in the text. The cycle length (expressed as a percentage of control cycle length) is plotted as a function of the phase in the cycle at which the stimulus is delivered (phase ranges from 0 at the start of the action potential to 100). The broken line represents no change in pacemaker length, and points along the diagonal represent "capture" with constant latency between stimulus and response. Two different current strengths were used. The control cycle length was 1575 msec. From Jalife and Moe (1976).

1000 msec following the action potential, the cycle length was decreased by about 18% less than control. These data are shown in figure 6.2, where the cycle length is plotted as a function of stimulus phase at two different stimulation strengths. Note the dependence of the cycle length on the phase of stimulus presentation.

Studies in many systems show the following generalizations concerning the effects of single perturbations on ongoing biological rhythms.

1. Following a perturbation, the rhythm is generally reestablished (following a transient) with the same frequency and amplitude as before the perturbation. The exception to this is oscillatory annihilation, discussed in section 5.4.

2. Although the rhythm is reestablished following a perturbation, its phase is shifted so subsequent marker events occur at times different from those which would have been observed in the absence of perturbation.

3. A single stimulus may lead to either a lengthening or a shortening of the perturbed cycle length, depending on the stimulus phase.

4. The graph of the cycle length as a function of the stimulus phase often has apparent discontinuities for some stimulus amplitudes, as illustrated in figures 6.1b and 6.2.

In the next two sections, we discuss these four experimental observations within the context of integrate and fire and limit cycle oscillator models, and show how the applicability of these two classes of models might be experimentally distinguished.

6.2 Phase Resetting in Integrate and Fire Models

Integrate and fire oscillator models (chapter 1) are popular in biology because they are conceptually simple, easy to analyze mathematically, and frequently give predictions consistent with experiment. To illustrate the phase-resetting predictions of integrate and fire models, we consider the respiratory rhythm.

A number of different workers have postulated respiratory integrate and fire models, with differing quantitative details. The essence of these models is contained in the highly oversimplified model shown in figure 6.3a. Normally the timing of respiration is assumed to be controlled by an activity oscillating between an inspiratory on-switch threshold and a inspiratory off-switch threshold. Inspiration is terminated when the inspiratory off-switch threshold is reached during the inspiratory phase. The signal to initiate inspiration is generated when the activity meets the inspiratory on-switch threshold, but inspiration does not start until after a brief delay.

The effect of a perturbation, which we assume to be a lung inflation, on this model oscillator depends on the phase of the cycle in which it is delivered. During inspiration, lung inflation transiently increases the activity by an amount S, proportional to the size of the inflation (figure 6.3b). If the inflation is sufficiently large, the inspiratory off-switch threshold is crossed and inspiration terminates. If the inflation is not large enough, then there is no effect on the inspiratory time. The graph of the added stimulus, S, required to terminate inspiration as a function of $1/T_I$ is shown in figure 6.4a, which should be compared with figure 6.1a. For lung inflations during expiration, we assume that the

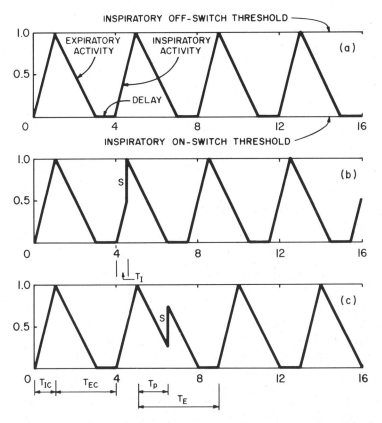

6.3. Integrate and fire model for respiratory rhythmogenesis. (a) Inspiratory activity rises to the inspiratory off-switch threshold. Then expiratory activity decreases to the inspiratory on-switch, and after a brief delay the cycle is repeated. The control inspiratory and expiratory times are T_{IC} and T_{EC}, respectively. (b) Effect of a stimulus S delivered during inspiration. (c) The same stimulus delivered during expiration.

activity is increased by an amount proportional to the inflation volume, and then relaxes in a linear fashion as before until the inspiration terminates (figure 6.3c). However, if the perturbation is delivered during the delay phase, there is no effect on the timing. Therefore the normalized expiration time, as a function of stimulus phase during expiration, should vary as in figure 6.4b (cf. figure 6.1b).

The graph of the normalized cycle time as a function of the phase of the stimulus presentation for this integrate and fire model is shown in figure 6.4c. There are three discontinuities, and the magnitude of these discontinuities increases with increasing stimulus strength. Furthermore, the position of the first discontinuity will move to the left,

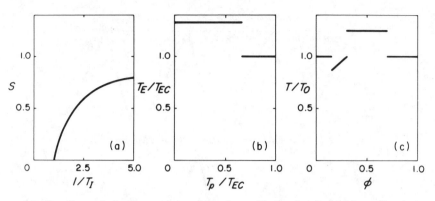

6.4. The effects of stimulus on the model in figure 6.3. Panels (a) and (b) are directly analogous to panels (a) and (b) of figure 6.1, and the symbols have the same meaning. (c) The *perturbed* cycle length as a function of the phase of the stimulus, where phase 0 is taken as the start of inspiration.

while the locations of the other two will remain stationary as stimulus strength is increased. Although experimental data that would allow a direct comparison with the predictions of figure 6.4c can easily be obtained they have, to our knowledge, not been reported.

This integrate and fire model is too simple to be taken seriously as a model for respiratory rhythmogenesis. However, similar models can be generated from this prototype by allowing nonlinear activities—for example, exponential decay of activity during expiration—and by allowing volume perturbations to affect the thresholds rather than the activities. In more detailed models like these, multiple discontinuities in the graph of cycle length versus phase of lung-inflation stimulus will also be observed.

In the natural sciences it is usual to find mathematical models for dynamical processes formulated as differential equations, rather than as integrate and fire models (see chapter 2). Although under some circumstances it may be possible to approximate a differential equation with an integrate and fire model, a careful analysis of the results of phase resetting in both types of models reveals important differences. In the next section we examine phase resetting of oscillations in nonlinear differential equations.

6.3 Phase Resetting of Limit Cycle Oscillations

Many mathematical models proposed for biological oscillations have stable limit cycles, for example, the models for nonlinear feedback, sequential disinhibition, and pacemaker cells described in chapter 4.

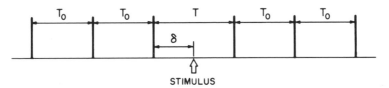

6.5. Experimental protocol for phase-resetting experiments. The perturbed cycle length T is determined as a function of the phase δ/T_0 of the stimulus, where T_0 is the control cycle length.

Interestingly, as Winfree has extensively documented, the responses of limit cycle oscillators to pulsatile stimuli display several general characteristics independent of the detailed mathematical equations describing the system. In this section we summarize the main results of phase resetting of limit cycle oscillators.

We first present the general experimental paradigm and definitions used in perturbation studies. The general situation is depicted in figure 6.5. The dark bars represent marker events in a spontaneous oscillation having an intrinsic or control period, T_0. These marker events might include, for example, the initiation of an action potential in a neural or cardiac preparation, the start of inspiration, or the start of mitosis in cells growing in tissue culture. The phase at the marker event is taken as 0. The *phase* at any subsequent time t, $0 < t < T_0$ is defined to be $\phi = t/T_0$. The phase as defined here lies between 0 and 1. In other notations the phase is expressed in degrees (multiply the phase by 360° or radians (multiply the phase by 2π). If a stimulus is delivered at a time δ following a marker event, the phase of delivery of the stimulus is δ/T_0.

In these perturbation experiments, the effect of the stimulus on the cycle in which it was delivered and on subsequent cycles is measured. The usual finding is that the duration of the cycle in which the stimulus is delivered is altered. This is called the *perturbed cycle length*, designated by T in figure 6.5. In addition, duration of subsequent cycles may be altered, but for simplicity we assume that they are not. This assumption is reasonable for many systems since it is often observed that there is a rapid return to the control cycle length following a perturbation. The basic experiment of delivering a perturbing stimulus is then repeated, varying both the stimulus phase and the magnitude.

As a result of the perturbation of the cycle in which the stimulus was delivered and the rapid approach to the normal cycle length, subsequent marker events occur at times that are different from those that would have been observed in the absence of perturbation. The difference in marker events is often reported as a normalized or relative

phase difference $\Delta\phi = (T - T_0)/T_0$. In some experimental preparations, the events immediately following the perturbation may be unob-
servable because of experimental artifact, so the phase difference can
be measured only after several periods of the autonomous cycle have
elapsed.

In this situation it is clear that the phase difference is ambiguous,
since phase differences of -0.1, 0.9, 1.9 would asymptotically appear
to be the same. This is a reflection of the fact that the asymptotic phase
shift is only measurable modulo 1. This ambiguity in the measurement
of phase shift can be confusing, and the phase-shift terminology (although often used) is best avoided unless it is explicitly clear how it is
defined. We will now assume that the cycle is generated by a limit cycle
oscillation. Let $x(t = 0)$ and $x'(t = 0)$ be the initial conditions of a point
on the cycle and a point not on the cycle, respectively, and $x(t)$ and
$x'(t)$ be the coordinates of the corresponding trajectories at time t. If
$\lim_{t\to\infty} d[x(t), x'(t)] = 0$, where d is the Euclidean distance, then the
latent phase of $x'(t = 0)$ is the same as the phase of $x(t = 0)$.

The locus of all points with the same latent phase is called an *isochron*. An isochron is a smooth curve (for limit cycles in two dimensions) crossing the trajectories in the attractor basin of the limit cycle.
The state point on any trajectory in the attractor basin of the limit
cycle passes through all the isochrons at uniform rate. Thus isochrons
are very close together wherever time derivatives are small. In particular, isochrons come arbitrarily close together at any fixed point and
therefore necessarily also along any singular trajectory leading to a
fixed point. The locus of stationary states and attracting sets of these
stationary states is called the *phaseless set*. Except for the phaseless set,
one and only one isochron passes through each point in the attractor
basin of the limit cycle. These concepts are illustrated in figure 6.6a,
which shows the isochrons for the Poincaré oscillator of equation (2.4)
and figure 2.3. In other, more complex differential equations, the isochrons will generally not be straight lines. The following discussion is
also applicable to these more complex situations.

The effect of a stimulus is to shift a state point on the limit cycle, at
some isochron ϕ, to a new point in phase space lying on some new
isochron ϕ' generally not on the limit cycle. Assume that the stimulus
is a horizontal translation by an amount b (figure 6.6b). Then, from
trigonometric arguments, ϕ and ϕ' are related by the relation

$$\cos 2\pi\phi' = \frac{b + \cos 2\pi\phi}{(1 + 2b \cos 2\pi\phi + b^2)^{1/2}}. \tag{6.1}$$

The function that can be used to compute the new phase following a
perturbation is called the *phase transition curve*, or PTC. The PTC is

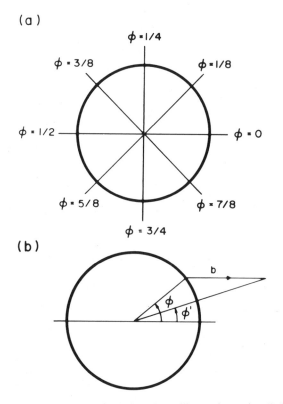

6.6. (a) The isochrons for the simple limit cycle oscillator of equation (2.4) and figure 2.3. (b) The effect of a perturbation. A stimulus delivered at phase ϕ resets the phase to ϕ', where ϕ and ϕ' are related by equation (6.1).

linked to the perturbed cycle length T by the formula,

$$\phi' = 1 + \phi - T/T_0, \qquad (6.2)$$

where we will take the intrinsic cycle length to be $T_0 = 1$.

A geometical interpretation of the phase resetting experiment is useful (figure 6.7). If stimuli are administered at all phases of the cycle, then the locus C' of new states reached immediately after the stimulus will be a displaced image of the limit cycle C. We call this closed curve, C', the *shifted cycle*. Phase changes continuously along C' except whenever C' cuts across the phaseless set. Assuming C' is continuous, one can circumnavigate C' once, while counting the net number of times ϕ' advances through a cycle (defined from the isochrons of the original oscillator). This integer is called the *winding number* or *topological degree* of C'. After a sufficiently slight perturbation (in this case $0 < b < 1$) the shifted cycle C' scarcely differs from the limit cycle C, so ϕ' scarcely

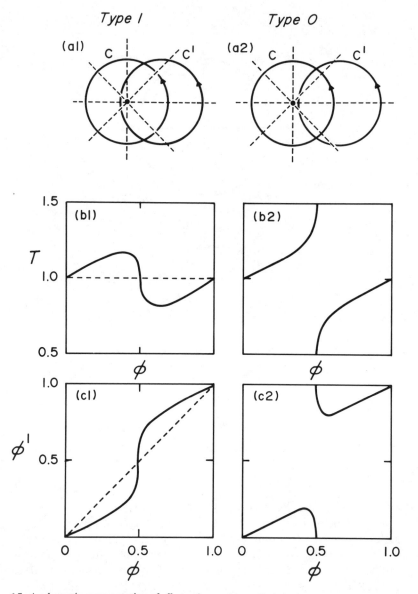

6.7. A schematic representation of effects of perturbing a limit cycle oscillator as in figure 6.6b. (a) Perturbations result in the shifted cycle C'. If the shifted cycle encloses the origin, there is type 1 phase resetting (a1), and if the shifted cycle does not enclose the origin, there is type 0 phase resetting (a2). (b) The perturbed cycle length for type 1 (b1) and type 0 (b2) phase resetting. (c) The new-phase, old-phase plots for type 1 (c1) and type 0 (c2) phase resetting. From Glass and Winfree (1984).

differs from ϕ, and the degree of C' is 1 (figure 6.7). We call this *type 1 phase resetting*. A perturbation that changes the winding number to 0 is said to induce *type 0 resetting*. An examination of figure 6.7 makes it clear that if the perturbation is sufficiently large ($b > 1$), type 0 phase resetting is expected.

Now consider the simple limit cycle in figure 6.6 as a conceptual model for the heartbeat. Assume that each time the phase passes through 0, this corresponds to the upstroke of the cardiac action potential (the marker event). The perturbation is assumed to be a depolarizing stimulus. These perturbed cycle-length data should be compared with the experimental findings in figure 6.2. There is a striking qualitative agreement between the perturbed cycle length in the simple model (figure 6.7) and the observed data in the phase-resetting experiment. This simple model is also consistent with the four main experimental observations on phase-resetting experiments outlined in section 6.1. Since experiments determine the new phase as a function of both stimulus phase and amplitude, the results can be displayed three-dimensionally. This three-dimensional helicoidal plot has been dubbed the *time crystal* by Winfree.

From a theoretical perspective, the analysis of the phase-resetting data based on the topological theory of limit cycles is simple and elegant. It serves to account qualitatively for the main experimental findings without the numerous ad hoc assumptions of integrate and fire models. In view of the compact nature and simplicity of the topological theory of limit cycle perturbation, we hope that sufficiently accurate data from other systems willl become available so perturbed cycle length and the PTC can be determined as a function of stimulus phase and amplitude.

Although the topological theory of limit cycle phase resetting qualitatively accounts for several features of phase-resetting experiments in a number of systems, there are sometimes difficulties when a detailed analysis is performed. We consider some of these in section 6.5. First, however, we examine phase resetting in various systems.

6.4 Phase Resetting in Diverse Systems

Phase-resetting experiments have been carried out in a large number of different experimental systems. Often, the researchers have not been familiar with the mathematical theory of phase resetting sketched above, and general underlying principles of phase resetting are not explicitly recognized, even though they are present. We give several examples of phase-resetting experiments in diverse systems.

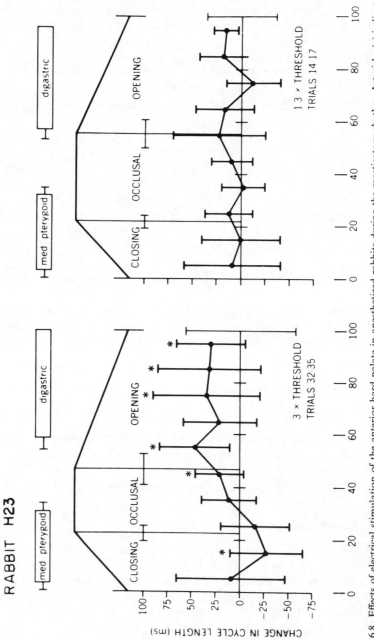

6.8. Effects of electrical stimulation of the anterior hard palate in anesthetized rabbits during the masticatory rhythm. Asterisks (*) indicate significant deviations from control ($P < 0.05$). From Lund, Rossignol, and Murakami (1981).

It has often been noted in studies of motor-rhythm generation that there are differential effects of stimuli delivered at different phases of the cycle. We have already discussed the effects of lung inflation and pointed out that the effects on the respiratory cycle are different if the inflation is delivered during inspiration and expiration. In similar fashion, electrical stimulation of the palate of anesthetized rabbits during mastication demonstrates that the effects of stimulation depend on both the amplitude and the phase of the stimulus (figure 6.8). The differential effects of electrical stimulation on reflexes during locomotion have also been well characterized, as we mentioned earlier. Although the details of the response and the underlying mechanisms need to be clarified, we must stress that the same stimulus delivered to any limit cycle model will generally have differential effects that depend on the phase in the cycle at which the stimulus is delivered.

In another interesting study of phase-resetting behavior, stimuli were delivered to patients who had either essential tremor or Parkinsonian tremor (figure 6.9a). The phase resetting was determined as a function of the phase in the cycle at which the stimulus was delivered. There was a large noise level in these experiments, but there were clearly strong phase-dependent effects on the ongoing rhythm. The amount of phase resetting was described by a linear function (figure 6.9b). The slope of this function reflected the extent to which the phase of the rhythm could be reset by peripheral stimuli and thus may help distinguish essential tremor from Parkinsonian tremor. Although such an approach is intriguing and of potential utility in assessing peripheral involvement in tremor, the linear fits to the data do not reflect what one expects for perturbation of limit cycle oscillators.

As a second clinical example of phase resetting, we consider the mechanism for the generation of cardiac rhythms with frequent ectopy. In the normal heart, the rhythm is set by a small region of specialized tissue called the sinoatrial (SA) node located in the right atrium. Many cardiac arrhythmias are believed to be due to the development of abnormal sites of pacemaker activity, called *ectopic pacemakers*, generating spontaneous rhythms that compete and interfere with the normal cardiac rhythm set by the SA node. This would lead to complex rhythms, called *modulated parasystole* by Jalife and Moe, resulting from the phase resetting of an independent ectopic focus by the normal sinus rhythm. If modulated parasystole prevails, it should be possible to demonstrate that a sinus beat falling at different phases of the ectopic cycle has differential effects. Consider the data shown in figure 6.10a in which the ectopic beats are labeled by X and the sinus beats by *R*. The inter-ectopic intervals during which a sinus beat fell could be

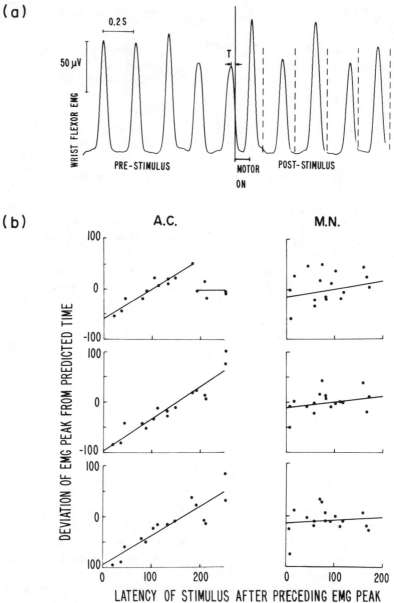

6.9. (a) Phase resetting of EMG bursts from the wrist flexors of a patient with essential tremor. The patient maintained a handle in a zone during the prestimulus period with a steady level of flexion against a torque motor. A brief displacement is delivered as indicated by the bar in the bottom of the figure. (b) Deviations in the timing of maximal EMG activity following a stimulus in two Parkinsonian patients. Statistically significant deviations, related to the timing of the stimulus, were observed during the tremor cycle in which the stimulus occurred (cycle 0; upper graphs) and two subsequent cycles (1 and 2; middle and lower graphs) for patient A. C. but not for patient M. N. Adapted from Stein, Lee, and Nichols (1978).

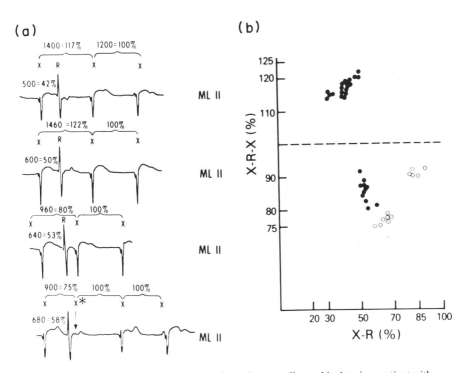

6.10. (a) Phase-resetting curves derived from electrocardiographic data in a patient with frequent ectopic beats. The ectopic beats are labeled by an X and the nonectopic beats by an R. A concealed ectopic discharge is labelled X*. All values are expressed in msec. (b) Phase-resetting curves relating the X-R-X intervals to the X-R intervals, both normalized as percentage of the ectopic cycle length (horizontal line at 100%). Open circles represent concealed discharges that can be inferred from the timing of the sybsequent ectopic discharges in the bottom panel in part (a). From Castellanos et al. (1984).

analyzed to give a phase-resetting curve (figure 6.10b) closely resembling that shown in figure 6.2.

These examples serve to illustrate that exactly the same experimental paradigm has been utilized by workers in several disciplines for very different reasons, though the underlying mathematical concepts involved are common.

6.5 Practical Problems with Application of the Topological Theory

Consider the simple nonlinear oscillator of figure 2.3 whose response to isolated stimuli was shown in figure 6.7. The steady state at $r = 0$ is unstable, so arbitrarily close perturbations to this point will diverge from it as time proceeds, and the oscillation will be reestablished. Consequently, even though there is a stimulus that will annihilate the oscillation, in any practical situation it would be impossible to deliver it exactly, and the oscillation would be reestablished. Also, in concrete situations the fluctuations present would send the dynamics away from unstable steady states. Thus, in practice, it is expected that there will exist situations in which it is impossible experimentally to annihilate oscillations by a single isolated stimulus, even though the oscillation is a limit cycle oscillation. As we discussed in section 5.4, if there is a stable steady state, then annihilation of an ongoing oscillation is possible.

Application of the topological theory presents practical difficulties. One of Winfree's basic ideas is to use the presence of type 1 phase resetting at low stimulus strengths and type 0 phase resetting at high stimulus strengths to infer information about phase resetting at intermediate strengths. The difficulties arise because without further experiments or a good deal of additional knowledge concerning the topological structure of the nonlinear oscillator, it is difficult to make accurate predictions about phase resetting at intermediate-strength stimuli. The following discussion is difficult, but it is important for all researchers who wish to apply the topological theory to concrete problems.

The key notion needed to apply the topological results to concrete situations comes from the following continuity properties of nonlinear limit cycle oscillations. If (1) a biological oscillator is generated by a stable limit cycle oscillation, and (2) all perturbations to the oscillation of fixed strength but at variable phases shift the current point in phase space to a second point that remains in the basin of attraction of the oscillation, then after any transients have died away the plot of new phase versus old phase for that fixed stimulus strength is a continuous function that maps the unit circle into itself (e.g., see ϕ' versus ϕ in figure 6.7).

In many situations, the plot of new phase versus old phase does not appear to be continuous. An example is provided from previously mentioned studies of phase resetting of aggregates of spontaneously beating cells from chick heart. At low-stimulus strengths, new phase–old phase was type 1, and at high stimulus strengths it was type 0. At intermediate strengths a different situation prevailed (figure 6.11). A sequence of stimuli delivered at 141 msec after an action potential led to a prolonged perturbed cycle length, and the same stimuli delivered at 143 msec led to a shortened perturbed cycle length. Stimuli delivered at 142 msec led to one of two responses—either a prolongation or a short-

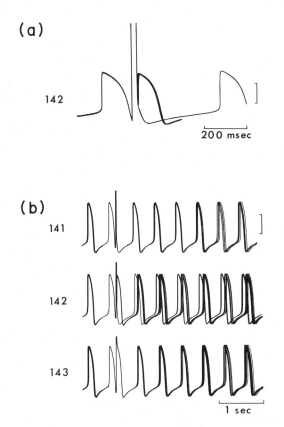

6.11. Phase-resetting data derived from spontaneously beating embryonic ventricular heart-cell aggregates. Current pulse stimuli were 27 nA amplitude and 20 msec duration. The times at the left of the figure indicate the coupling interval from the start of the action potential to the stimulus in msec. This shows a discontinuous response (see discussion in the text). From Guevara, Shrier, and Glass (1986).

ening of perturbed cycle length. Letting the oscillation continue for several cycles, it is clear that the envelopes of action-potential wave forms following prolongation and shortening are distinct (i.e., they do not superimpose asymptotically as they must if the new phase versus old phase curve were continuous).

The data of figure 6.11 pose a fundamental problem. No perturbations were found which shifted the oscillation out of its basin of attraction at the stimulus strength delivered. Consequently, if the phase-resetting function is discontinuous, as it appears to be, then the basic rhythm could not be generated by a limit cycle oscillation. The differential equations that model ionic current flow through physiological membranes suggest that, although the phase-resetting curves appear to be discontinuous, they may in fact be continuous. Simulations have shown that these ionic models can give extremely steep phase-resetting functions that would be difficult to resolve experimentally. Experimental tests of the continuity properties that require the resolution of time increments at less than a picosecond accuracy—as the simulations would suggest is necessary—are impossible because of noise and equipment limitations. Thus for all practical purposes, experimentally measured phase-resetting functions will appear to be discontinuous provided the model predictions are as accurate in this situation as they have proved to be in many others. If the PTC is discontinuous at some stimulus strengths, then topological arguments based on continuity of this function cannot be invoked. Therefore, the use of continuity arguments without the explicit demonstration of continuity of the PTC function is always questionable.

Along the same lines, it is important to recall that the continuous phenomenological ionic models represent averages of fluctuating currents through ionic channels. Consequently, theoretical interpretations based on continuous differential equations can be valid only if time and current steps are sufficiently large so that fluctuations in individual channels do not play a role. At the very fine increments that have been used to discuss phase resetting in ionic models, channel fluctuations are important and may give rise to apparent discontinuities in experiments.

6.6 Summary

In most situations, a single-pulse perturbation delivered to a spontaneously oscillating physiological system will act to reset the phase of the ongoing rhythm. The magnitude of the resetting depends on both the stimulus magnitude and the phase of the stimulus in the cycle. The graph of the new phase as a function of the old phase (the phase transition curve or PTC) is either a continuous function with winding number 1 (type 1) or 0 (type 0), or it is discontinuous. In general, integrate and fire models give discontinuous PTCs, and limit cycle models give type 1 PTCs at low-stimulus strengths and type 0 PTCs at high-stimulus strengths. In many experimental systems and mathematical models, rapid changes in the PTC over narrow ranges of stimulus phase make an unambiguous identification of the qualitative features of the PTC difficult, if not impossible.

An implicit assumption of much theoretical work has been that a differential equation is the most appropriate model for an oscillating biological phenomenon. The experimentally observed discontinuities, which are hard to account for using models formulated as differential equations, do occur in integrate and fire models (section 6.2). Under certain special circumstances, integrate and fire models represent limiting cases of limit cycle oscillators and often provide a simple conceptual model for rhythmogenesis in physiological systems.

Notes and References, Chapter 6

6.1 Overview of Experimental Results

The vast literature dealing with the phase resetting of biological rhythms as a result of perturbations is extensively discussed in Winfree (1980, 1987a,b).

For the original description of the Hering-Breuer reflex, see Breuer (1868). Adrian (1933) demonstrated the role of the vagal afferents in the modulation of the reflex in a classical paper. Further studies of the role of vagal afferents in respiratory physiology have been carried out by many others, and we have used the important results of Clark and Euler (1972) and Knox (1973). Iscoe and Vanner (1980) have studied the long time resetting of the respiratory rhythm due to vagal stimulation. For a current review of respiratory rhythmogenesis that includes vagal reflex mechanisms, see Feldman (1986).

Phase resetting in the Purkinje fiber preparation is described in Jalife and Moe (1976, 1979) and reviewed in Jalife and Michaels (1985).

6.2 Phase Resetting in Integrate and Fire Models

Many workers have proposed integrate and fire models for respiratory activity, and the papers of Bradley et al. (1975), Herczynski and Karczewski (1976), Remmers (1976), Cohen and Feldman (1977), Baconnier et al. (1983), and Petrillo and Glass (1984) are representative. Equally many others have proposed nonlinear differential equations with limit cycle behavior as models for the central respiratory oscillator (Feldman and Cowan 1975; Geman and Miller 1976; Fincham and Liassides 1978; Pham Dinh et al. 1983; Baconnier et al. 1983). In certain limits, a limit cycle oscillator can be approximated by an integrate and fire model. This point was made initially in a discussion of the control of mitosis by Tyson and Sachsenmaier (1978). Another example of this correspondence, applicable to respiration, is in Petrillo and Glass (1984).

6.3 Phase Resetting of Limit Cycle Oscillators

The main references for the theory of phase resetting of limit cycles are the books and articles by Winfree (1975, 1977, 1980, 1987a,b), which should be consulted for extensive historical information, examples, and further references. For other theoretical discussions of phase resetting of limit cycle oscillations, see Pavlidis (1973), Guckenheimer (1975), Kawato and Suzuki (1978), and Kawato (1981). The presentation in the text follows along the lines of Glass and Winfree (1984).

Van Meerwijk et al. (1984) applied the theory to experiments on the phase resetting of the cardiac rhythm in aggregates of cells from embryonic chick heart. It is also possible that the theory is applicable to respiratory-rhythm generation. A study on the effects of superior laryngeal nerve stimulation on the respiratory rhythm in cats gave evidence for type 1 phase resetting at low stimulus strength and type 0 phase resetting at higher stimulus strengths (Paydarfar, Eldridge, and Kiley 1986). Similar results were found in a study in which electrical stimulation was delivered directly to the brain (Paydarfar and Eldridge 1987).

6.4 Phase Resetting in Diverse Systems

Phase-resetting experiments have been done in a very large number of other systems, as previously noted. Of particular interest are studies on mastication (J. P. Lund, Rossignol, and Murakami 1981) and locomotion (Forssberg et al. 1976). Phase resetting as a diagnostic tool in Parkinsonism has been employed by R. B. Stein, Lee, and Nichols (1978) and Lee and R. B. Stein (1981). The analysis of modulated parasystolic rhythms based on phase resetting has been discussed by Jalife and Moe (1976, 1979) and Moe et al. (1977). Clinical applications of these ideas have been considered by Furuse, Matsuo and Saigusa (1981), Jalife, Antzelevitch, and Moe, (1982), Nau et al. (1982), and Castellanos et al. (1984). A good recent review is Jalife and Michaels (1985). Additional discussion of the theory of modulated parasystole is in section 7.5.

6.5 Practical Problems with Application of
   the Topological Theory

The topological theory for phase resetting sketched out in section 6.3 is subtle, and unfortunately it has been frequently misinterpreted in the experimental literature. For example, Jalife and Antzelevitch (1979) state: "His [Winfree's] theory, in which techniques of differential topology are used, predicts that if the phase resetting of the pacemaker in response to the perturbation follows certain specific patterns [i.e., it is type 1 at low stimulus strength and type 0 at high stimulus strength], then there must be a characteristic stimulus magnitude and timing at which pacemaker activity is completely annihilated." Even though this reasoning led Jalife and Antzelevitch (1979) to discover annihilation of action potentials with a single stimulus, the statement is nevertheless misleading since there can be an unstable phase singularity. In this case, the oscillation would always be reestablished following a perturbation.

The numerical characteristics of the phase resetting predicted by Hodgkin-Huxley-like differential equations that model excitable membrane behavior have been given by Best (1979), Chay and Lee (1984, 1985), and Clay, Guevara, and Shrier (1984). These studies show that observation of the topological properties of the phase-resetting functions (e.g., type 1 or type 0 behavior) is expected to be difficult in practice because of the very steep slopes of the PTC. These analyses offer a possible explanation for the apparently discontinuous phase-resetting behavior observed in Guevara, Shrier, and Glass (1986). Theoretical arguments which assume continuity of the PTC can be found in Winfree (1983b, 1987b).

Chapter 7

Periodic Stimulation
of Biological Oscillators

$\mathbf{P}$eriodic stimulation of spontaneously oscillating physiological rhythms has powerful effects on the intrinsic rhythm. As the frequency and amplitude of the periodic stimulus are varied, a variety of different coupling patterns are set up between the stimulus and the spontaneous oscillator. In some situations the spontaneous rhythm is entrained or phase locked to the forcing stimulus so that for each N cycles of the stimulus there are M cycles of the spontaneous rhythm, and the spontaneous oscillation occurs at fixed phase (or phases) of the periodic stimulus ($N:M$ *phase locking*). In addition to phase-locked rhythms, it is also possible to observe irregular or aperiodic rhythms in which fixed phase relationships and regular repeating cyclic patterns are not observed. In section 7.1 we discuss the main experimental observations in phase locking experiments, with particular reference to experiments on the cardiac and respiratory systems. In section 7.2 we develop the mathematical concepts needed to analyze phase locking theoretically. These techniques are applied to analyze phase locking in integrate and fire models in section 7.3 and in limit cycle models in section 7.4. In section 7.5 we discuss several phenomena entailing phase locking in humans.

7.1 Overview of Experimental Results

One experimental paradigm for studying physiological oscillators is to subject the oscillator to periodic stimuli while maintaining physiological conditions as constant as possible. We do not attempt to give a complete summary of this large body of work but discuss two illustrative systems: the mechanical ventilation of animals, and the periodic stimulation of cardiac oscillations using an intracellular microelectrode. We draw generalizations from these systems which are broadly applicable in a wide variety of experimental systems.

It has been known since the time of Hering and Breuer that periodic lung inflation by a mechanical ventilator can lead to phase locking between the ventilator and the intrinsic respiratory rhythm in mammals. The entrainment is believed to be mediated by the Hering-Breuer reflexes in which expansion of the lungs inhibits inspiration and prolongs expiration (see chapter 6). Afferent activity from stretch receptors in the lungs is carried by afferent fibers in the vagus nerve. Respiratory entrainment can be studied by mechanically ventilating an animal at different ventilator volumes and frequencies while attempting to maintain constant levels of anesthesia, body temperature, and blood gases.

Experiments were performed on paralyzed, pentobarbital-anesthetized adult cats, and central respiratory activity was monitored by recording from a branch of the phrenic nerve. In a normal unparalyzed animal, the phrenic nerve innervates the diaphragm, and phrenic nerve acitivity causes the diaphragm to contract (lower), thus leading to inspiration. In the paralyzed animal, neuromuscular transmission between the phrenic nerve and the diaphragm is blocked, and lung inflation is due solely to the mechanical ventilator. However, afferent activity from stretch receptors in the lung is still carried by the vagus nerve. Special steps have to be taken to maintain blood gases constant as the frequency and amplitude of the ventilator are varied. If the animal is hyperventilated, central respiratory activity is lost due to low CO_2. One method for maintaining constant blood gases is to vary the frequency and volume of the ventilator simultaneously, so that the total ventilation per unit time is maintained constant. However, this technique cannot be used if one wishes to obtain information about the ventilator-respiratory rhythm coupling as a function of both frequency and volume (varied independently). In this case, an alternative method is to add CO_2 to the inspired gas in order to maintain constant physiological levels of CO_2 even at high ventilator volumes and frequencies.

As the ventilator volume and frequency are varied, a number of different rhythms are established between the ventilator and phrenic activity. These different rhythms are organized in an orderly fashion in the ventilator volume–ventilator frequency plane shown in figure 7.1. Insets show representative traces of ventilator volumes and phrenic nerve activity corresponding to different stable phase-locked rhythms, as well as non-phase-locked rhythms (which occur in the shaded regions). In these experiments, low ventilation frequencies and volumes could not be studied since adequate ventilation must be maintained. Likewise, very high volumes and frequencies could not be studied because of the mechanical limitations of the ventilator and the limited lung capacity of the cat.

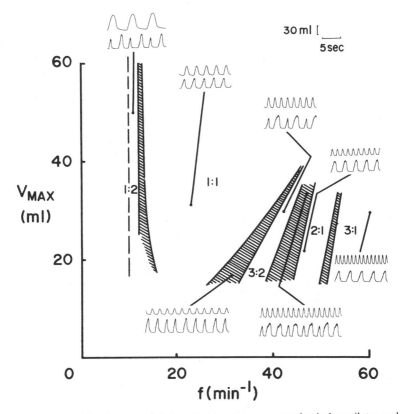

7.1. Composite showing zones of phase locking between a mechanical ventilator and the respiratory rhythm obtained experimentally in anesthetized, paralyzed cats. In each inset the upper trace is lung volume and the lower trace is integrated phrenic activity. The scales are the same for all insets. The shaded regions represent irregular dynamics. V_{max} is the maximum ventilator volume and f is the ventilator frequency. From Petrillo and Glass (1984).

As discussed in section 4.2, it is believed that the respiratory rhythm is generated in a complex network of neurons in the brain stem. In contrast, the cardiac rhythm is generated in a specialized region of electrically coupled cells in the SA node that act as a pacemaker for the heart. We now describe experiments in which the effects of pulsatile electrical stimuli delivered to spontaneously beating cells derived from the ventricles of embryonic chick heart were determined.

As the frequency and current intensity of the electrical stimuli are varied, a variety of different rhythms between the stimulator and the heart cells are established. The results of these experiments are summarized in the composite in figure 7.2. In this figure the insets represent

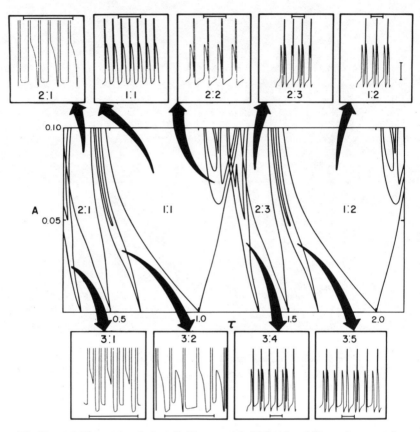

7.2. Theoretically computed phase-locking zones (solid lines) and illustrative traces (insets) for periodically stimulated aggregates of embryonic chick-heart cells. *A* is stimulus amplitude (arbitrary units) and τ is the period of the stimulus divided by the intrinsic cycle length. Calibration bars are 1 sec and 50 mv. From Glass, Guevara, and Shrier (1987).

the different observed phase-locking patterns, and the solid lines represent the results of theoretical computations based on phase-resetting experiments using single pulses. In addition to the stable phase-locked rhythms, there are a number of irregular rhythms. In section 7.4 we discuss the theoretical techniques used to compute the phase-locking zones and associate the observed irregular rhythms with chaotic dynamics.

Although these results have been obtained from two very different physiological systems using different types of periodic stimulation, there are certain striking features common to both. The following generalizations are applicable to a large number of experiments of periodic forcing of biological oscillators.

1. The stable zones of phase locking most commonly observed correspond to low-order ratios between the number of cycles of the forcing stimulus and the intrinsic rhythm (i.e., $2:1$, $3:2$, $1:1$, $2:3$, $1:2$). Although other $N:M$ ratios with larger values of N and M can also be observed, these occupy smaller areas in the frequency-amplitude parameter space, and they are consequently easily overlooked or obscured by noise.

2. The stable rhythms are organized in the frequency-amplitude plane in an orderly fashion. It is common to associate a *rotation number* $\rho = M/N$ with an $N:M$ rhythm. Then, as the stimulation frequency increases at fixed stimulus amplitude, ρ decreases.

3. At very low stimulation amplitudes, it is difficult to maintain stable phase locking.

4. If the regions of frequency-amplitude parameter space between stable phase-locking zones are studied, then it is generally possible to find stimulation parameters that give rise to irregular dynamics.

Despite their similarities, there are differences between the cardiac and respiratory systems. For example, the $2:2$ region observed in the periodically stimulated cardiac cells was not observed in the mechanically ventilated cats. The goal of theoretical studies of phase locking is to provide a theoretical basis for understanding the similarities and differences between the different preparations. Ideally, one would like to be able to make predictions about the phase locking as stimulation parameters vary, based on the mechanisms of rhythmogenesis and the coupling of the stimulator to the intrinsic rhythm. In practice, the mathematical analysis of periodically forced nonlinear oscillators is an extremely difficult problem, and detailed quantitative understanding of dynamics has been obtained in only a few special situations.

The theoretical studies make it clear that although there are many similarities between the dynamics in different systems, there will also generally be differences if the dynamics are studied in sufficient detail. In order to establish this assertion, we describe the dynamics resulting from periodic stimulation in a number of different model systems. We also discuss entrainment of biological oscillators in a number of clinical contexts.

7.2 Mathematical Concepts

The study of periodically forced nonlinear oscillators has a rich history, and it is still an area of active research. Here we give only the most important results from the perspective of experimental studies in biology.

Periodically forced nonlinear oscillators were studied in the 1920s by van der Pol and van der Mark. They proposed that the electrical activity of the heart could be modeled by three nonlinear oscillators corresponding to the sinus node, the atria, and the ventricles. There is a unidirectional coupling between the sinus and the atrial oscillators, and likewise a unidirectional coupling between the atrial and ventricular oscillators. By reducing the coupling strength between the atrial and ventricular oscillators, they found it was possible to obtain a number of different stable phase-locked rhythms that correspond qualitatively to a class of cardiac arrhythmias called the *atrioventricular (AV) heart blocks*. However, most workers in cardiovascular physiology attribute AV heart block to blocked conduction in the AV node, rather than to lack of synchronization between atrial and ventricular oscillators (see section 8.1).

The simple two-dimensional differential equation proposed by van der Pol to model nonlinear limit cycle oscillations has played an important role in applied mathematics. Studies of the effects of periodic sinusoidal forcing of this equation were undertaken by van der Pol and still continue. The periodically forced van der Pol equation can be written as

$$\frac{d^2u}{dt^2} - \varepsilon(1 - u^2)\frac{du}{dt} + u = B\cos(vt). \tag{7.1}$$

When $B = 0$, there is a unique stable limit cycle oscillation. As v and B vary, there are entrainment regions, as shown in figure 7.3. A notable observation made first by Cartwright and Littlewood and subsequently by Levinson in the 1940s was that it is possible to find parameter values such that aperiodic orbits exist for some set of initial conditions. In the current jargon of nonlinear dynamics, these aperiodic orbits correspond to chaotic dynamics in the periodically forced nonlinear oscillators. In a celebrated paper published in 1967, Smale, who was following Levinson's exposition, discovered a mathematical construction (the *horseshoe map*) that shows the existence of an infinite number of periodic orbits, as well as aperiodic orbits in a special class of two-dimensional finite difference equations (diffeomorphisms of the plane).

As an introduction to the mathematical analysis of periodically forced nonlinear oscillators, consider the hypothetical situation in which a biological oscillator is periodically stimulated. To be concrete, imagine the forcing to be a sinusoidal electrical-current perturbation of a spontaneously oscillating neuron. Take the period of oscillation of the sine wave to be 1 sec, and assume that the period of the oscillation of the nerve cell (T_0) is different from 1 sec. Keep a record of the sequence of times that the nerve cell fires.

In the limit of no coupling between the neuron and the perturbation, the sinusoidal forcing has no effect. In this limit it is a trivial problem

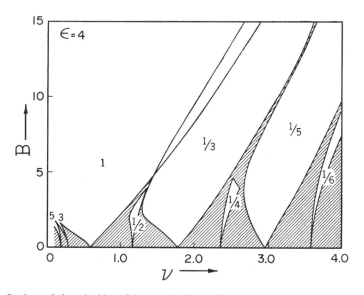

7.3. Regions of phase locking of the van der Pol oscillator (equation 7.1) to a sinusoidal stimulation obtained using an analog computer. The ratios between the cycles of the forced oscillator and the forcing function are indicated. Dynamics in shaded regions were called "beat oscillations" (now called quasiperiodicity). From Hayashi (1964).

to compute the next firing time of the neural oscillator. If t_i is the time of the ith firing, then

$$t_{i+1} = t_i + T_0. \tag{7.2}$$

If we are only interested in the phase of the sinusoidal stimulus at which the neuron fires, then it suffices to consider only the fractional part of the values of t_i. Mathematically, we take t_i modulo 1 and call the resulting phase of the ith firing ϕ_i (figure 7.4a). For example, if we take $T_0 = 0.75$ sec, then $\phi_{i+1} = \phi_i + 0.75$ (mod 1). A graph of this function is shown in figure 7.4b. In the more general situation with coupling, the sinusoidal stimulus will have an effect on the firing of the neuron. However, if the coupling is comparatively weak, there may only be small changes from the zero-coupling limit. In this case, equation (7.2) can be rewritten

$$\phi_{i+1} = \phi_i + f(\phi_i, b) + T_0 \qquad \text{(mod 1)}, \tag{7.3}$$

where the function $f(\phi_i, b)$ is (generally) a nonlinear function that depends on the coupling strength b. In the limit of zero coupling, $f(\phi_i, 0) = 0$.

Equation (7.3) is another example of a finite difference equation (see section 2.5). Once the nonlinear function f is known, it is possible to

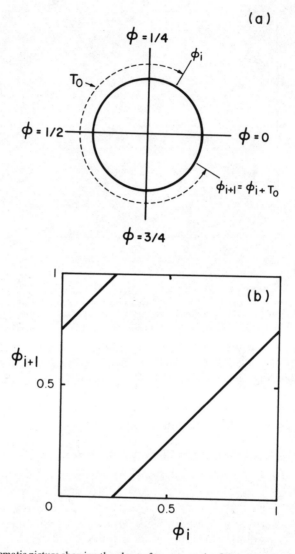

7.4. (a) Schematic picture showing the phase of two successive firings of a neural oscillator in a sinusoidal cycle assuming there is no influence of one upon the other. (b) A graph of the associated finite difference equation.

compute the dynamics for all future times. However, in contrast to the quadratic map (equation 2.6), equation (7.3) contains two parameters, b and T_0. The firing times t_i (mod 1) can be represented as points on the circumference of a circle of unit circumference. Then the iteration of equation (7.3) takes one point on the circle to a second point. Such

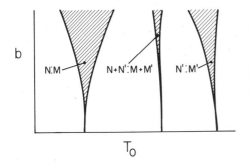

b

$N{:}M$ $N{+}N'{:}M{+}M'$ $N'{:}M'$

T_0

7.5. Schematic diagram of Arnold tongues. In the shaded regions there is a stable phase locking between a spontaneously oscillating system and an imposed periodic input function. Between any two stable phase locking zones there are always others. The ordinate represents the amplitude and the abscissa the period of a periodic forcing function.

a function is called a *circle map*. To understand the effects of perturbations, it is important to understand the changes in dynamics (bifurcations) that arise as parameters in the circle map are varied.

Provided the nonlinear function f in equation (7.3) is not too large, the bifurcations of this equation are well understood. The precise mathematical definition of "not too large" is that there must be a 1:1 correspondence between values of ϕ and ϕ'. Thus for any value of ϕ there is one and only one possible value of ϕ', and for each value of ϕ' there is one and only one value of ϕ. Such maps are called *invertible*.

Analysis of the bifurcations of invertible circle maps was undertaken by Poincaré in the last century and has remained a topic of great interest. Major advances were made by the Russian mathematician Arnold. The results of his analysis are schematically shown in figure 7.5. The (b, T_0) plane has distinct regions, called *Arnold tongues* or *Arnold horns*, which correspond to stable phase locking in a ratio $N{:}M$ (N cycles of the stimulator and M cycles of the neural oscillator). Arnold tongues are present for all rational ratios $N{:}M$ where N and M are relatively prime integers (i.e., they have no common divisor). This means that there are an infinite number of Arnold tongues corresponding to all possible ratios between the frequencies of the stimulator and the driven oscillator. It should be clear that the basic structure shown in figure 7.5 is also present in figures 7.2 and 7.3, at least over some of the range of parameter space.

In invertible circle maps, for all parameter values inside the Arnold tongue corresponding to $N{:}M$ phase locking, all initial conditions asymptotically approach a stable $N{:}M$ phase-locking pattern. We informally define the *rotation number* as the ratio between the number of cycles of the stimulator divided by the number of cycles of the forced oscillator. Thus for $N{:}M$ phase locking, the rotation number $\rho = M/N$ (see also the Mathematical Appendix).

Are there any combinations of stimulus amplitude and frequency that do not give rise to stable entrainment? Said another way, what dynamics are found between Arnold tongues? There exist parameter values for which no stable entrainment is found for any initial condition. The dynamics in this situation are called *quasiperiodic*. The easiest way to imagine this is to consider the case in which the two rhythms are completely independent. As time proceeds, the phase relations between the two rhythms will continually shift but, in the general case, will never repeat if the ratio between the two frequencies is not rational. If every value of ϕ_i which is generated is plotted on the circumference of a circle, then, in the limit as $i \to \infty$, the circumference will be densely covered with points (this means that any point on the circumference is arbitrarily close to a point in the sequence $\phi_0, \phi_1, \phi_2, \ldots$). The dynamics are aperiodic, but they are not chaotic since two initial conditions that are close together remain close together in subsequent iterations.

These dynamics are sometimes called *oscillatory free runs* or *relative coordination*. An example showing a typical appearance for quasiperiodic dynamics is given in figure 7.1 in the inset showing the dynamics present between the 1:1 and 3:2 rhythms. Although one might expect that it would be hard to find sets of parameter values that give rise to quasiperiodicity, there is a finite probability that one will observe quasiperiodicity given a random choice of parameters in invertible circle maps. In fact, in experimental studies at low stimulation strength, the usual experience is that dynamics appear to display quasiperiodicity rather than phase locking.

The discussion to this point has considered a situation in which the coupling strength (and hence the nonlinearity) is not too large and the dynamics are represented by invertible circle maps. In concrete experimental systems and in mathematical models, as the strength of the periodic perturbation increases it is no longer possible to represent the dynamics by invertible circle maps, and the comparatively simple and well-understood Arnold tongue structure in figure 7.5 is no longer observed. We now consider several ways in which the Arnold tongue structure can be destroyed by considering the periodic forcing of integrate and fire and limit cycle oscillations.

7.3 Periodic Forcing of Integrate and Fire Models

Periodic modulation has been incorporated into integrate and fire models in two different ways. First, we may assume that the activity is periodically modulated but the threshold is held constant. Alterna-

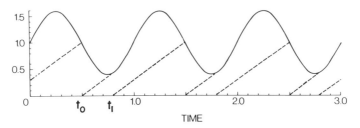

7.6. Integrate and fire model with a sinusoidally modulated threshold. After reaching the threshold there is an instantaneous reset to zero. Starting at an initial condition t_0, the threshold will be reached the first time at t_1, which can be calculated using equation (7.5). 1:2 phase locking is shown. Adapted from Glass and Mackey (1979b).

tively, the threshold may be subjected to periodic modulation. In the following we discuss the effects of threshold modulation.

One motivation for the analysis of periodic forcing of integrate and fire models comes from the experimental studies on the mechanical ventilation of cats. An examination of the insets in figure 7.1, in particular the 1:2 and 3:2 rhythms, shows large lung inflations coincident with the premature termination of the inspiratory activity. This gives a striking visual suggestion for the utility of an integrate and fire model with modulated threshold for this system. As we already discussed in chapter 6, the Hering-Breuer reflexes can be modeled by integrate and fire models.

We consider a simple integrate and fire model with a periodically modulated threshold (figure 7.6). For reasons of concreteness and simplicity, we assume that the threshold is the sine function

$$\theta(t) = 1 + k\sin(2\pi t). \tag{7.4}$$

We assume the activity increases linearly, with slope γ, to the threshold and then resets to zero. The goal is to understand in detail the bifurcations and dynamics in this model as a function of the parameters for all initial conditions.

Assume that t_0 is known (figure 7.6). Then the time the activity first reaches threshold t_1 can be found by solving the equation,

$$\gamma(t_1 - t_0) = 1 + k\sin(2\pi t_1). \tag{7.5}$$

This is a transcendental equation in t_1 and has no analytical solution. However, t_1 in equation (7.5) is defined implicitly by a finite difference equation and can be numerically computed once the initial condition and the parameters are specified. Some of the entrainment zones are shown in figure 7.7.

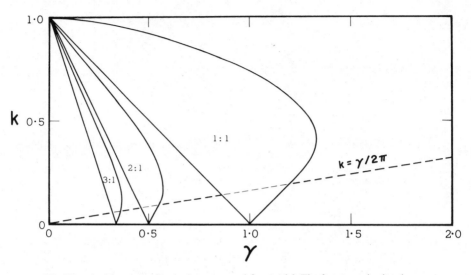

7.7. Phase-locking zones for the integrate and fire model. The frequency in the absence of sinusoidal modulation ($k = 0$) is $f = \gamma$. The Arnold tongue structure is present beneath the dashed line. From Glass and Bélair (1986).

For $k < \gamma/2\pi$ the dynamics are described by a continuous invertible map of the unit circle, and the discussion above concerning Arnold tongues is applicable. If, however, the slope of the rising activity is less than the maximum slope of the sine wave, then the dynamics will be described by a discontinuous noninvertible function. In this case, it is known that there will still be $N:M$ phase locking for all integers N and M that are relatively prime. There are also parameter values that give rise to aperiodic dynamics. There are, however, two differences between the properties of the aperiodic dynamics in the case in which the dynamics are described by the discontinuous piecewise monotonic maps and by the invertible maps. First, successive iterates no longer form a dense orbit on the unit circle. A dense orbit is clearly impossible since there is a forbidden range of values of t_{i+1} for which no pre-image exists. Second, the probability that one will choose a set of parameter values associated with the aperiodic dynamics is now zero.

In addition to studies of integrate and fire models in which the thresholds are sinusoidal, there have been several studies of the properties of integrate and fire models in which the thresholds are piecewise linear. Such studies allow a much more complete analysis of the dynamics than is possible using sinusoidal thresholds. For example, Lasota and Mackey proved that such models display chaotic dynamics in some parameter ranges. The significance of this finding lies in the observation of chaos even in extremely simplified models for the periodic forcing of oscillations. In view of this finding, it seems likely that the appearance of chaotic dynamics at least over some range of stimulation parameters in periodically forced physiological oscillators may be very common.

Attempts to develop more realistic integrate and fire models for the entrainment of the respiratory rhythm have been made by Petrillo and Glass. They assume that two thresholds are modulated by the volume of the mechanical ventilator. The timing of inspiration and expiration is represented by activities that oscillate between the two thresholds. A brief delay occurs between the time the inspiratory onset threshold

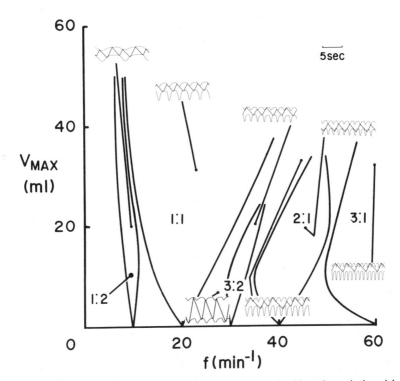

7.8. Composite showing different zones of phase locking obtained in a theoretical model for mechanical ventilation of a paralyzed, anesthetized cat. In the model, which is displayed in the insets, there are two thresholds that are modulated by the periodic lung inflations (not shown). The solid line in the insets represents the operation of the respiratory rhythm generator that rises and falls linearly between the thresholds. Compare with figure 7.1. From Petrillo and Glass (1984).

was reached and the beginning of inspiration. This model has five parameters, all of which could be determined from the data. Numerical simulations of the model (figure 7.8) show good agreement with the experimental results shown in figure 7.1. However, a detailed study of the bifurcations in this model was not carried through.

Another attempt to develop integrate and fire models for the entrainment of biological rhythms was made by Daan and coworkers for the circadian rhythm. They assumed that there were two sinusoidally modulated thresholds. With this model they were able to provide a partial explanation for data on sleep duration as a function of the time of onset of sleep. A systematic analysis was not made of the zones of 1:1 entrainment (corresponding to the normal circadian rhythm), or of other possible phase-locking zones as a function of the parameters.

In conclusion, the assumption that there are periodic inputs to integrate and fire models provides a conceptually simple means to model phase locking in biological systems. Even the simplest possible models lead to extremely complicated dynamics, which are only partially understood.

7.4 Entrainment of Limit Cycle Oscillators

Many biological rhythms are best represented mathematically as limit cycle oscillations in differential equations (see chapter 4). Because these rhythms interact with each other and because there is periodic stimulation from the external environment, it is important to understand the effects of periodic forcing on a limit cycle oscillation. One prototypical model for periodically forced limit cycles is the sinusoidally forced van der Pol equation (see equation 7.1 and the related discussion). J. H. Jensen and coworkers have shown that sinusoidal forcing also gives rise to chaotic dynamics in mathematical models of excitable neural and cardiac tissue, and Aihara and coworkers have demonstrated strange attractors from sinusoidally forced squid axons (see section 3.4).

We now consider the effects of a periodic train of short pulsatile stimuli on limit cycle oscillations. In the event that the limit cycle is rapidly reestablished following a single stimulus, it is straightforward to compute the effects of periodic stimuli once the effects of a single stimulus are understood.

The main idea can be developed from a consideration of the effects of periodic stimulation of the Poincaré oscillator considered in chapters 2 and 6. Once again assume that there is a limit cycle at $r = 1$, that perturbations consist of a horizontal translation by an amount b, and that following a perturbation the limit cycle is rapidly approached. As discussed in chapter 6, the effect of a single stimulus delivered at phase ϕ is to shift the limit cycle to a new phase, ϕ', where

$$\phi' = g(\phi, b). \tag{7.6}$$

The function g is called the phase transition curve (PTC). If ϕ_i is the phase immediately preceding the ith stimulus, then the phase immediately preceding the $(i + 1)st$ stimulus is simply

$$\phi_{i+1} = g(\phi_i, b) + \tau \quad \text{(mod 1)}, \tag{7.7}$$

where τ is the time interval between periodic stimuli measured relative to the intrinsic cycle length of the limit cycle oscillator (figure 7.9). Equation (7.7), which is equivalent to equation (7.3), is the necessary relationship; equation (7.7) is a general result. If the PTC can be computed or experimentally measured, and if there is a rapid return to the limit cycle following a perturbation, then once an initial condition is chosen the dynamics can be determined by iteration for all future times.

For the simple limit cycle model the PTC can be readily computed, and analytic and numerical techniques can be applied to determine the detailed structure of the phase-locking zones as a function of stimulus amplitude and frequency. In this example, when $b < 1$ the PTC is an invertible type 1 circle map, and for $b > 1$ the PTC is a type 0 circle map. Although the bifurcations that result are not completely known, this is perhaps the best understood nontrivial example of the effects of periodic forcing on a limit cycle oscillation over a broad range of

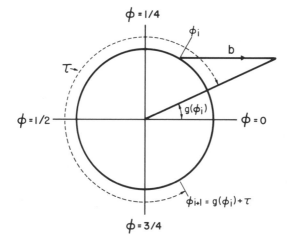

7.9. A schematic model for the perturbation of a limit cycle by a periodic stimulus. Provided there is a rapid relaxation back to the limit cycle, equation (7.8) is derived.

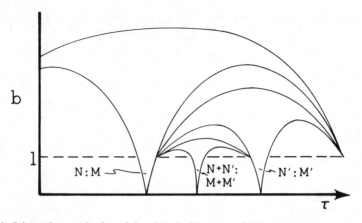

7.10. Schematic organization of the phase-locking zones for the periodic stimulation of the nonlinear oscillator in figure (7.9), assuming rapid relaxation back to the limit cycle following each stimulus. Beneath the dashed line is the Arnold tongue structure. Above it there are complex bifurcations, as discussed in Guevara and Glass (1982), Hoppensteadt and Keener (1982), and Keener and Glass (1984). From Glass and Bélair (1986).

stimulus frequencies and amplitudes. The results are summarized in figure 7.10, which shows a schematic representation of the continuation of the Arnold tongues. This model also shows chaotic dynamics and complex bifurcations over some regions of parameter space. In this model the extensions of some of the Arnold tongues that do not extend to high amplitudes have a mushroom shape. In fact, in both the cardiac and the respiratory experiments it was impossible to find some of the more complex rhythms (such as 3:2 entrainment) at the high values of stimulation intensity.

The application of the results in this section to the study of concrete systems can be carried out in the following way. From figure 6.6b it is easy to see that the effects of a single pulse delivered to the oscillator at a phase ϕ will result in a perturbed cycle length,

$$T/T_0 = 1 + \phi - g(\phi), \tag{7.8}$$

where $g(\phi)$ is the PTC. Thus, by experimentally measuring the perturbed cycle length as a function of ϕ, it is possible to measure experimentally the PTC, which can then be numerically iterated to determine the entrainment zones. This procedure has been carried out on the periodically stimulated cardiac cells, and the solid lines of figure 7.2 were computed in this manner. There was close agreement between the theoretical calculations and the experimental observations.

The main assumption in these calculations is that the relaxation back to the limit cycle is sufficiently fast so only a single variable, the phase

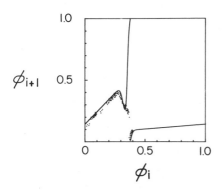

7.11. A plot showing the phases of successive stimuli as a function of the preceding stimulus, as in figure 1.11. The solid superimposed line is derived from single-pulse-perturbation experiments using the theory in the text. The failure of points to superimpose (as they should) in the region $\phi_i = 0.38$ is probably due to inaccuracies of the fitted curve to the phase resetting data in this region. From Glass Shrier, and Bélair (1986).

of the current stimulus, need be considered. An a posteriori justification for this assumption is provided by the determination of ϕ' versus ϕ directly from experimental tracings during aperiodic (chaotic) dynamics, and from comparison of this with the same curve derived from equation (7.6) using experimentally measured values of $g(\phi)$. This is illustrated in figure 7.11, where the superimposition of the points on the solid curve (determined from single-pulse perturbation experiments) demonstrates that the one-dimensional models presented here are appropriate for the description of the experimental cardiac dynamics. It is also important to recognize that nonmonotonic circle maps can lead, at some parameter values, to chaotic dynamics similar to those observed in the simple quadratic map described in chapter 2. Despite the success of the one-dimensional theory, an important experimental problem is to determine the effects of prior stimulation history on the current state of the system. At some point it will be necessary to consider higher-dimensional finite difference equations.

7.5 Phase Locking of Rhythms in Humans

The theoretical work described above concentrated on the different phase-locking patterns and rhythms observed from the periodic forcing of very simple models of biological oscillators. Even in these simple systems, the detailed behavior under periodic stimulation is so complex that it is unlikely it will ever be observed in any system, let alone in a biological system that tends to be "noisy." Despite the difficulty for fine observations of predicted behavior, an understanding of the gross phenomenology may have potential significance in a very large number of normal and pathophysiological situations in humans. We present a number of examples. In no case is there a theory that is well worked out.

Respiratory Sinus Arrhythmia

This phenomenon refers to the modulation of the cardiac rate by respiration, due to complex coupling between these two rhythms. During the inspiratory phase there is lowered intrathoracic pressure, which leads to greater cardiac filling and hence higher stroke volumes. Blood-pressure differences resulting from the changes in stroke volume will lead to differences in baroreceptor reflexes, which in turn lead to different afferent vagal effects on the cardiac cycle. Differences in oxygenation of the blood during inspiration and expiration will lead to different chemoreceptor reflexes at the different phases of the respiratory cycle. These, in turn, may affect respiratory and cardiac rhythms. Finally, the brain-stem activity associated with respiratory rhythmogenesis can also lead to fluctuations in sympathetic and vagal tone and therefore affect cardiac control. Despite—or perhaps because of—these multiple feedback loops, in humans there is comparatively weak coupling between respiration and the cardiac rhythm, and the resulting rhythms are generally not phase locked. Rather, a slight acceleration of the cardiac rhythm is observed during inspiration, and a slight slowing down is observed during expiration, giving rise to a rhythm that appears to be quasiperiodic.

Respiratory-Locomotory Coupling and Interlimb Coordination

Butler and Woakes showed that during locomotion in birds there are complex biomechanical interactions between the musculoskeletal elements involved in generating the locomotory and respiratory rhythms. In quadrapedal (four-footed) mammals there is normally a 1:1 entrainment between the respiratory and locomotory rhythms. Bramble and Carrier have shown that in a galloping horse, 1:1 entrainment at rates of over 100/min can be observed. In running humans, the coupling between respiration and locomotion can be studied by recording breathing sounds (pneumosonograms) and foot impact during locomotion. A number of different coupling patterns have been observed (figure 7.12), but a complete understanding of the reflex mechanisms underlying the entrainment is not available.

Ectopic Cardiac Pacemakers

We have already discussed the resetting of ectopic pacemakers by the normal sinus rhythm. A related problem is to predict the rhythms resulting from the interaction of sinus and ectopic pacemakers based

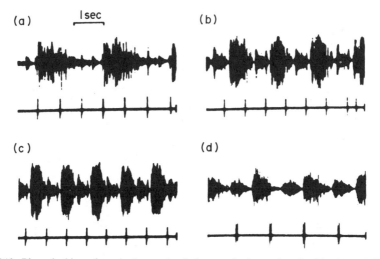

7.12. Phase locking of respiration and gait from a single run in a healthy human. In each panel the upper trace records the sound produced by respiration and the lower trace indicates the impact of the right foot. (a) 3:1 phase locking; (b) 2:1 phase locking; (c) 3:2 phase locking (running); (d) 3:2 phase locking (walking). From Bramble (1983).

on the phase-resetting curves. A calculation along these lines was undertaken by Moe and coworkers (figure 7.13). The various ratios (3:2, 2:1, 5:2, etc.) labeled at the top of the figure show the zones of stable entrainment. However, because of the refractory periods of the heart, not all ectopic beats will be observed. The apparently complex organization of the zones similar to those observed in periodically stimulated heart cells (figure 7.2) is as expected. Although the clinical condition in which there is ECG evidence of only a single ectopic focus with ectopic beats occurring less frequently than about 6/min is currently considered to be relatively benign, arrhythmias due to multiple ectopic foci are potentially dangerous and can degenerate to tachycardia and fibrillation (see section 8.5). Indeed, in a classic text in electrocardiography published in 1946, Katz used to term "chaotic heart action" to describe complex arrhythmias in which there were multiple ectopic foci. The connection between these rhythms and the "chaotic" dynamics that are the focus of this book remains to be clarified.

Ventilator-Respiration Coupling

In many acute and chronic clinical settings, mechanical ventilation is necessary. Sometimes it is difficult to adjust the ventilator so that the patient does not "fight the ventilator." When this happens, there

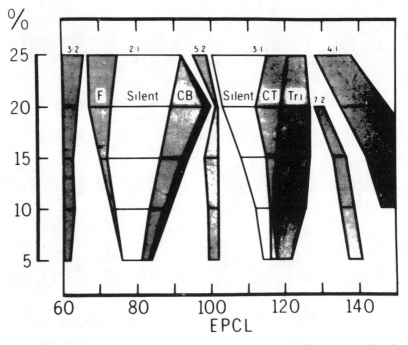

7.13. Phase-locking zones in a mathematical model of parasystole. The sinus cycle length is 40 and the ectopic cycle length (EPCL) varies as shown on the abscissa. The ordinate is a measure of the strength of the influence of the sinus pacemaker on the ectopic rhythm. The zones give stable entrainment between the sinus rhythm and the ectopic pacemaker, but not all ectopic beats are observed because some fall during the refractory period of the ventricles. In the silent zones, all ectopic beats fall in the refractory period, and in the concealed bigeminy (CB) and concealed trigeminy (CT) zones only some of the beats fall in the refractory period. In the region labeled F, there are fusion beats that arise when the ectopic pacemaker and sinus beat fall at approximately the same time. From Moe et al. (1977).

are several strategies: sedate, paralyze or hyperventilate the patient. Obviously, it is most desirable to adjust the ventilator so that there is some reasonable entrainment pattern with patient and ventilator infla- tions in phase. A systematic study of the entrainment of the respiratory rhythm in humans has been carried out. The subjects were anesthetized, and diaphragmatic activity was monitored by an esophageal electrode during ventilation with a fixed volume without patient triggering. It was possible to demonstrate 1:1 entrainment over a range of ventilator frequencies and amplitudes (figures 7.14, 7.15). This situation is com- plicated in many currently available ventilators, since it is possible to have ventilatory efforts by the patient triggering the inflation pulse of the ventilator.

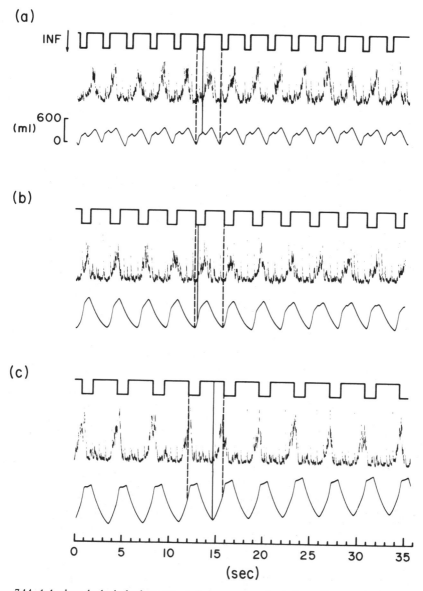

7.14. 1:1 phase-locked rhythms set up between a mechanical ventilator and the respiratory rhythm in an anesthetized human subject. In each panel the top trace is the
mechanical ventilator, the second line is the integrated diaphragmatic electromyogram
recorded with an esophageal electrode, and the third line is the ventilation volume.
Dashed and solid lines show onset of mechanical ventilator and diaphragmatic inspiratory efforts, respectively. (a) ventilator frequency (f) = 24/min, ventilator volume (V) =
400 ml. (b) f = 19.4/min, V = 500 ml. (c) f = 15.6/min, V = 600 ml. From Graves et al.
(1986).

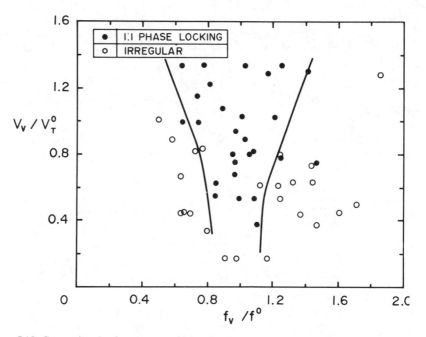

7.15. Composite plot from seven subjects showing the region of 1:1 phase locking between a mechanical ventilator and the respiratory rhythm in mechanically ventilated anesthetized humans. The abscissa is the ventilator frequency divided by the intrinsic frequency in the absence of the mechanical ventilator, and the ordinate is the ventilator volume divided by the intrinsic tidal volume. An approximate boundary separating the 1:1 zone from zones of irregular dynamics has been drawn. From Graves et al. (1986).

Sleep Arrhythmias

Many physiological variables display a 24-hour periodicity, arising from an entrainment of the sleep-wake and other physiological cycles to the 24-hour day. However, as Wever has shown, if an individual lives in a constant environment in which social and physical clues of the external environment are not present, then an intrinsic circadian rhythm develops which differs from the normal 24-hour rhythm.

Many psychiatric patients with affective disorders show abnormalities in their circadian rhythms. For example, many depressed patients awake early and complain about their sleep habits. Practitioners are currently suggesting that interventions directed toward manipulation of the circadian rhythm may be useful therapeutically. Thus Wirz-Justice is studying the effects of pharmacological manipulations, whereas Lewy and coworkers have been manipulating the light-dark cycle.

On a more mundane level, most of us have experienced jet lag, and many of us have experimented with various sleeping, eating, or pharmacological strategies to minimize it. Since controlled experiments are difficult, we are sure to witness a great deal of quackery in the years ahead in this area. Formulation of models is not straightforward because to date it has been impossible to measure the parameters in the models directly or to manipulate them.

7.6 Summary

In response to periodic stimulation, physiological rhythms may become either entrained (phase locked) to the stimulus giving periodic dynamics, or the resulting rhythms can be aperiodic. In the case of aperiodic rhythms, quasiperiodic or free-running rhythms are observed at small-stimulus amplitudes. At higher-stimulus amplitudes, aperiodic dynamics can sometimes be associated with chaos. The organization of the entrainment zones as a function of the frequency and amplitude shows a number of striking features. At low-stimulus amplitude, an Arnold tongue structure is observed. In this case one expects that if there is $N:M$ entrainment at one frequency and $N':M'$ entrainment at another, then an intermediate frequency can be found at which there is $N + N':M + M'$ entrainment. However, the range of frequencies over which such dynamics will be observed may be so small that, in practice, experimental observations are impossible. As stimulus amplitude increases, the Arnold tongue structure breaks down and may give a complex topological structure. Entrainment phenomena are believed to be important in physiology and occur in a number of different systems.

Notes and References, Chapter 7

7.1 Overview of Experimental Results

A number of workers have carried out systematic studies of the entrainment of respiration to a mechanical ventilator in a number of different mammalian preparations (Fallert and Muhlemann 1971; Vibert, Caille, and Segundo 1981; Baconnier et al. 1983). The results we report, taken from Petrillo, Glass, and Trippenbach (1983) and Petrillo and Glass (1984) are in agreement with results from these other laboratories.

Likewise, a number of workers have studied the effects of periodic electrical stimulation of cardiac pacemaker tissue, either using sinusoïdal or pulsatile stimuli (Reid 1969; Levy, Iano, and Zieske 1972; Van der Tweel, Meijler, and Van Capelle 1973; Jalife and Moe 1976, 1979; Scott 1979; Ypey et al. 1982; Jalife and Michaels 1985). The experimental results given here on the periodic stimulation of embryonic heart-cell aggregates are extracted from an extensive series of experimental and theoretical studies (Guevara, Glass, and Shrier 1981; Glass et al. 1983; Glass et al. 1984; Guevara 1984; Guevara, Shrier, and Glass 1988).

The generalizations concerning the periodic forcing of biological oscillators are supported by a number of studies in diverse systems (Perkel et al. 1964; Pittendrigh 1965; Pavlidis 1973; Pinsker 1977; Ayers and Selverston 1979; Guttman, Feldman, and Jakobsson 1980).

7.2 Mathematical Concepts

It is a difficult problem to analyze mathematically the dynamics of periodically forced nonlinear oscillations such as the van der Pol equation (van der Pol 1926; van der Pol and van der Mark 1928). For representative work, see Cartwright and Littlewood (1945), Levinson (1949), Hayashi (1964), Flaherty and Hoppensteadt (1978), Levi (1981), and Guckenheimer and Holmes (1983). Our understanding of this problem is still far from complete, and it will certainly remain a rich area for mathematical research in the future.

A basic understanding of the dynamics of invertible circle maps is due to Poincaré (1885, 1954), Denjoy (1932), and Arnold (1965), and a good summary for the mathematically sophisticated reader is found in Arnold (1983) and Devaney (1986). The analysis of circle maps is currently of great interest and is discussed in more detail in the Mathematical Appendix.

7.3 Periodic Forcing of Integrate and Fire Models

An early application of integrate and fire models was to study the response of sensory systems to cyclic input (Rescigno et al. 1970; Knight 1972; Fohlmeister, Poppele, and Purple 1974). A comprehensive recent review of the dynamics in integrate and fire models in the presence of cyclic input is given in Keener, Hoppensteadt, and Rinzel (1981).

Early studies analyzed the integrate and fire model in figure 7.6 in the context of periodically forced relaxation oscillations (Harker 1938; Builder and Roberts 1939). Unaware of these early studies, the same model was considered in the context of entrainment of biological oscillators (Glass and Mackey 1979b; Glass et al. 1980). This model has also been proposed by Winfree (1980) in the context of circadian rhythms. Further analysis of mathematical aspects of the bifurcations in this model is in Keener (1980), Keener (1981), and Keener, Hoppensteadt, and Rinzel (1981). Modifications of this model are discussed in Glass and Bélair (1986).

Piecewise linear integrate and fire models have been investigated by several workers. Allen (1983) and Bélair (1986) have been able to derive precise boundaries for phase-locking zones. Lasota and Mackey (1985) have been able to prove the existence of chaotic (exact) dynamics.

Integrate and fire models for respiration are given in Baconnier et al. (1983) and Petrillo and Glass (1984). Integrate and fire models for circadian rhythms are presented in Winfree (1980, 1983a, 1984), Daan and Beersma (1984), and Daan, Beersma, and Borbely (1984) and are discussed by Strogatz (1986).

7.4 Entrainment of Limit Cycle Oscillators

A discussion of chaos resulting from sinusoidal forcing of nonlinear cardiac and neural oscillators is in J. H. Jensen et al. (1983), J. H. Jensen, Christiansen, and Scott (1984), and Aihara et al. (1986).

The use of finite difference equations to determine the effects of periodic stimulation of nonlinear oscillators with brief stimuli was originally presented in studies on the periodic forcing of neural oscillators (Perkel et al. 1964) and circadian oscillators (Pittendrigh 1965). Subsequently, equivalent procedures have been used in the analysis of diverse systems. For representative papers, see Keller (1967), Pavlidis (1973), Moe et al. (1977), Zaslavsky (1978), Scott (1979), Pinsker (1977), Ikeda, Tsuruta, and Sato (1981), Guevara, Glass, and Shrier (1981), Segundo and Kohn (1981), Ypey et al. (1982), Glass et al. (1983), Guevara et al. (1983), Honerkamp (1983), Guevara (1984), and Glass et al. (1984).

The study of the dynamics in the periodically forced Poincaré oscillator was undertaken in Guevara and Glass (1982), Hoppensteadt and Keener (1982), and Keener and Glass (1984).

7.5 Phase Locking of Rhythms in Humans

Respiratory sinus arrhythmia. This is a well known clinical phenomenon (for example, see Bellett 1971).

Respiratory-locomotory coupling and interlimb coordination. The coordination between fins in swimming was studied in a classic and beautiful work by von Holst (1973), who also noted analogies between fin arrhythmias and cardiac arrhythmias. Applications of the theoretical concepts to study limb coordination in turtles is reviewed by P.S G. Stein (1977). Extensive studies of respiratory-locomotory coordination have been carried out in birds (Butler and Woakes 1980) and mammals (Bramble 198., Bramble and Carrier 1983).

Ectopic cardiac pace takers. The model of modulated parasystole, as formulated by Jalife and Moe (1976, 1979) and Moe et al. (1977), has been extremely influential in both theoretical studies (Ikeda, Tsuruta, and Sato 1981; Honerkamp 1983) and clinical studies (Castellanos et al. 1984). Reviews of this topic can be found in Jalife and Michaels (1985) and Glass et al. (1987b).

Ventilator-respiration coupling. Despite its practical clinical importance, this phenomenon has been rarely studied in human patients (Curzi-Dascalova et al. 1979; Graves et al. 1986).

Sleep arrhythmias. A number of different mathematical models for circadian-rhythm generation and the entrainment of the circadian rhythm based on integrate and fire models (Winfree 1980, 1983a, 1984; Daan and Beersma 1984; Daan, Beersma, and Borbely 1984) and limit cycle models (Wever 1979; Gander et al. 1984) have been proposed (see the collection by Moore-Ede and Czeisler 1984). A comprehensive review of sleep disorders is in Weitzman (1981), and an extensive collection of papers on circadian rhythms in psychiatry has appeared (Wehr and Goodwin 1983). Particularly interesting applications of theoretical ideas to clinical work are in Kripke (1983), Wirz-Justice (1983), and Lewy, Sack, and Singer (1985). A reported phase resetting of the circadian rhythm by bright light is in Czeisler et al. (1986). Phase resetting of circadian rhythms with tranquilizers in hamsters has recently been reported (Turek and Losee-Olsen 1986).

Chapter 8

Spatial Oscillations

Physiological rhythms are ordered in space as well as in time. In the usual circumstance, the oscillation will display simple wavelike propagation originating from a pacemaker region. However, in many situations simple periodic wavelike propagation originating from a point source is not found, and other types of spatial ordering are observed. In this chapter we discuss these unusual modes of wave propagation. Wave propagation in a one-dimensional strip of tissue is considered in section 8.1 and in a one-dimensional ring of tissue in section 8.2. In section 8.3 we discuss wave propagation in two dimensions, and in section 8.4 we discuss wave propagation in three dimensions. Fibrillation is believed to be associated with abnormal reentry or circus spread of excitation in two and three dimensions. Theoretical and experimental studies of fibrillation are discussed in section 8.5.

8.1 Wave Propagation in One Dimension

A prototypical example of a physiological system in which there is one-dimensional wave propagation is the ureter. The ureters transport urine from the kidney to the bladder via peristaltic waves originating in a localized region near the junction of the ureter and the renal pelvis called the renal pelvic pacemaker. Wave propagation in the ureter can be studied either by measuring spontaneous activity or by stimulating one end of the ureter electrically and measuring electrical activity distally. In data from a study on intact dogs (figure 8.1a), a 1:1 transmission of excitation at a stimulus period of 4 sec (upper trace) and a 5:4 transmission of excitation at a stimulus period of 3.38 sec (lower trace) was observed. There is an increasing latency prolongation in the 5:4 rhythm between the stimulus and the next excitation wave, until eventually there is a nonconducted beat.

An analogous effect occurs in impulse conduction in the human heart. Even though the heart is a complex three-dimensional structure, for the current purposes we think of it as one-dimensional, with excitation

(a)

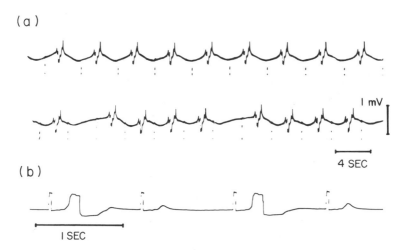

I mV

4 SEC

(b)

I SEC

8.1. Response of the ureter to periodic electrical stimulation. (a) Recording from anesthetized dog. The dashed line represents the stimulus and the continuous curve the activity recorded from an extraluminal bipolar electrode. The top trace shows 1:1 conduction when the stimulation period was 4 sec and the bottom trace shows Wenckebach cycles when the stimulation period was 3.38 sec. From Weiss, Wagner, and Hoffman (1968). (b) Recording from rat ureter in vitro. Electrical activity shows alternation of large and small responses from periodic stimulation. From Prosser, Smith, and Melton (1955).

following the path: SA node–atria–AV node–His bundle–Purkinje fibers–ventricles. Contraction of the atria is usually followed 0.12–0.20 sec later by contraction of the ventricles (figure 1.2). However, AV heart block is characterized by abnormal coordination between the atrial and ventricular rhythms, leading to a prolonged interval between atrial and ventricular contractions (first-degree AV block), more atrial than ventricular contractions resulting from blocked conduction of some atrial beats (second-degree AV block), or complete lack of coordination between atrial and ventricular rhythms (third-degree AV block). Second-degree heart block, in which there is an increasing interval between the atrial and ventricular contractions (the PR interval), leading to a dropped beat, is shown in figure 8.2a. Rhythms with increasing latency between stimulus and response and an eventual dropped beat are now called *Wenckebach rhythms*.

Another phenomenon, which occurs in both the ureter and the heart, is an alternation of responses to a periodic train of stimuli. Figure 8.1b shows an alternation of response that can be observed by rapid periodic stimuli delivered to an in vitro rat-ureter preparation. Similar phenomena are found in cardiology when alternation of pulse strength or electrical complexes (*alternans*) show up in electrocardiograms (figure

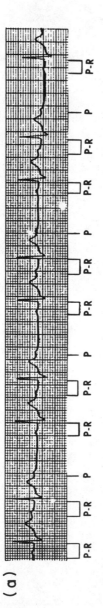

(a)

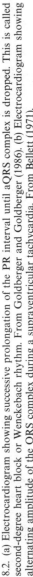

(b)

8.2. (a) Electrocardiogram showing successive prolongation of the PR interval until a QRS complex is dropped. This is called second-degree heart block or Wenckebach rhythm. From Goldberger and Goldberger (1986). (b) Electrocardiogram showing alternating amplitude of the QRS complex during a supraventricular tachycardia. From Bellett (1971).

8.2b). At rapid heart rates, alternans is frequently found and can be considered a normal response, but the appearance of alternans at lower heart rates is often considered a sign of dysfunction. Although an alternans response to periodic stimulation of one-dimensional excitable tissue at high stimulation rates is not encountered as often as Wenckebach rhythms, the response is sufficiently common to warrant a discussion of a possible mechanism for it.

In summary, periodic stimulation of excitable one-dimensional tissue at low rates leads to periodic traveling excitation waves of equal amplitude following the periodic stimulus in a 1:1 fashion. As the frequency is increased, the simple 1:1 propagation of equal-amplitude waves is no longer observed. One possibility is that each stimulus still leads to a wave of excitation, but the amplitudes of subsequent waves vary. A second possibility is that some of the waves are blocked and that periodic rhythms of the form $N:M$ with $N > M$ are found. Although more complex patterns of propagation are also clearly possible, we confine our attention to these two main types of dynamical behavior.

The physiological basis for understanding the dynamics arises from the following three observations, known to Mines early in the century: (1) As the stimulation frequency increases, the propagation velocity of a wave decreases (the dependence of propagation speed on frequency is called a *dispersion relation* in physics); (2) as the stimulation frequency increases, the duration of the excitation decreases; and (3) following an excitation there is a time interval, the *refractory period* (θ), during which a second stimulus cannot lead to a subsequent excitation. Incorporating these physiological properties into a mathematical model gives partial insight into the dynamics described above.

A schematic representation of the physiological situation is shown in figure 8.3. A stimulus (S_i) leads to excitation after a delay (SR_i). The

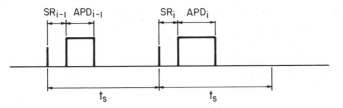

8.3. Schematic diagram showing the periodic stimulation of excitable tissue with a time interval of t_s between successive stimuli. SR_i represents the time interval from the ith stimulus to the start of the next action potential, and APD_i is the duration of the ith action potential.

action-potential duration (APD) of the excitation following stimulus S_i is denoted APD_i. The stimulus and the resulting excitation wave are often measured at different positions in the system being studied. Although in realistic situations SR_i and APD_i may depend on the stimulation history, we assume that the duration of these intervals is a function of the time from the end of the preceding excitation to the stimulus S_i. Calling t_s the time interval between stimuli, we have

$$SR_i = F(t_s - SR_{i-1} - APD_{i-1}) \qquad (8.1a)$$

$$APD_i = G(t_s - SR_{i-1} - APD_{i-1}), \qquad (8.1b)$$

where F and G are functions to be determined. Although equations (8.1a,b) have been introduced in the context of wave propagation in distributed systems, they are also applicable to spatially homogeneous systems in which stimulus and response are measured at the same point. To the best of our knowledge, a general analysis of equations (8.1a,b) has not been given. However, two special cases have been treated.

In the first case, we assume that the duration of excitation is constant and that, consequently,

$$SR_i = F(t_s - SR_{i-1}). \qquad (8.2)$$

In cardiac electrophysiology, the function F has been studied in the context of propagation of action potentials through the AV node. F is called the *recovery curve*. In this case, the interval SR corresponds from the beginning of the P wave on the ECG to the beginning of the QRS complex (figure 1.2). Since the early years of this century, it has been recognized that the PR interval is a decreasing function of the preceding RP interval, and that the AV recovery curve can be used to compute the effects of periodic stimulation as a function of the frequency of the periodic stimuli.

To illustrate the use of the recovery curve in practical situations, consider the following. An intra-atrial electrode is used to periodically stimulate a human heart, and the activity in the bundle of His was recorded. The interval between the stimulus and the onset of activity in the bundle of His is the SH interval. At a stimulation period of 440 msec, a Wenckebach block was established with 11 atrial contractions to 10 ventricular contractions (11:10 AV block) (figure 8.4a). Successive measurements of SH as a function of the preceding HS are shown in figure 8.4b. The data can be fitted to the exponential function,

$$SH = SH_{\min} + \alpha_1 e^{-HS/\tau_1} \qquad \text{for } HS > \theta, \qquad (8.3)$$

where $SH_{min} = 230$ msec, $\alpha_1 = 308$ msec, $\tau_1 = 111$ msec, and $\theta = 50$ msec. Using the recovery curve of equation (8.3), we can now derive

$$SH_i = SH_{min} + \alpha_1 e^{-(Nt_s - SH_{i-1})/\tau_1},\tag{8.4}$$

where N is the smallest integer, such that $Nt_s - SH_{i-1} > \theta$.

By iterating equation (8.4) it is possible to determine the dynamics for any value of t_s. A graphical iteration for $t_s = 442$ msec gives rise to a $10:9$ Wenckebach rhythm (figure 8.4c). In figure 8.4d we show the ratio between the number of conducted beats divided by the number of stimuli, ρ, as a function of t_s. The figure is the graph of a *Cantor function*. Since there are an infinite number of steps, most of unimaginably small size, the graph of a Cantor function is sometimes called a *devil's staircase*.

The results of figure 8.4d are quite general. With any monotonically decreasing recovery curve like that shown in figure 8.4b, conduction will be described by the Cantor function. Thus we observe the classical Wenckebach rhythms of the form $N:N-1$ as well as more complex variants. In this fashion the iteration of the finite difference equation serves to give a unified perspective and classification for Wenckebach and related arrhythmias based on the bifurcations predicted from the recovery curve.

We now consider equation (8.1) in a second situation, in which the time SR_i remains constant, but the APD varies. Specifically, we consider the effects of periodic stimulation on a spatially homogeneous excitable aggregate of cells from embryonic chick heart that is not spontaneously beating. Figure 8.5a shows the response of this preparation to periodic excitation with $t_s = 180$ msec. There is an alternation of APD which is not present at stimulation frequencies with $t_s > 200$ msec. The function that gives the APD as a function of the time from the end of the previous action potential to the stimulus is called the *electrical restitution curve*. The restitution curve can be approximated by the exponential function,

$$APD = APD_{max} - \alpha_2 e^{-\lambda/\tau_2},\tag{8.5}$$

where APD_{max} is the maximum APD, λ is the time that elapsed from the end of the preceding action potential to the stimulus, and α_2 and τ_2 are positive constants. The graph of the electrical restitution curve is shown in figure 8.5b. Substituting equation (8.5) into equation (8.1) gives

$$APD_i = APD_{max} - \alpha_2 e^{-(Nt_s - APD_{i-1})/\tau_2},\tag{8.6}$$

where N is the smallest integer such that $Nt_s - APD_{i-1} > \theta$. In equation (8.6) a period-doubling bifurcation will arise provided $\tau_2 \ln(\alpha_2/\tau_2) > \theta$

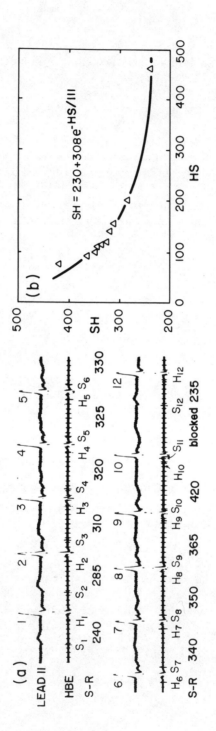

(b)

$SH = 230 + 308e^{-HS/III}$

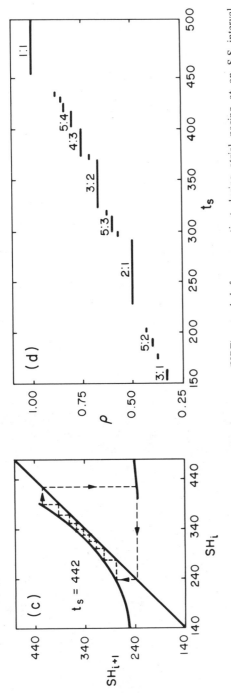

8.4. (a) Electrocardiographic lead II and His-bundle electrogram (HBE) recorded from a patient during atrial pacing at an S-S interval of 440 msec. S and H present the stimulus artifact and His-bundle deflections on the HBE, respectively. The time interval from each stimulus to the next His-bundle deflection is given in msec. From Levy et al. (1974a). (b) A plot of the SH interval (time from S to following H deflection) as a function of the preceding HS interval for ; data in (a). (c) Equation (8.4) with $t_s = 442$ msec. A cycle corresponding to 10:9 AV block is shown. The conduction ratio, ρ, (number of ventricular contractions divided by the number of atrial stimuli) as a function of t_s computed from equation (8.4) with the parameters in (b). This curve is called a Cantor function (Mandelbrot 1977, 1982). From Glass, Guevara, and Shrier (1987a).

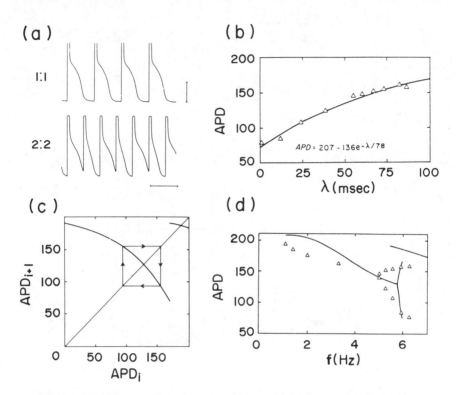

8.5. (a) Intracellular recording of transmembrane potential from a periodically stimulated quiescent heart-cell aggregate. In the upper trace the period of the stimulation is 300 msec, and in the lower trace it is 180 msec. Vertical calibration is 50 mv and horizontal calibration is 300 msec. (b) Action potential duration (APD) as a function of recovery time λ. (c) Equation (8.6) for $t_s = 170$ msec. There is a stable cycle with APD alternating between 94 msec and 156 msec (2:2 rhythm) and a stable steady state with APD = 187 msec (2:1 rhythm). (d) Bifurcation diagram showing APD as a function of stimulation frequency f. Solid lines show theoretical results, and triangles give data points. Adapted from Guevara et al. (1984).

at a critical stimulation frequency f^*, where $1/f^* = APD_{max} - \tau_2 + \tau_2 \ln (\alpha_2/\tau_2)$. At the critical simulation frequency the APD is $APD_{max} - \tau_2$ and the recovery time is $\tau_2 \ln(\alpha_2/\tau_2)$. Thus ın this experimental system we identify alternans with a period-doubling bifurcation. A graphical solution of equation (8.6) is shown in figure 8.5c for $t_s = 170$ msec. For this stimulation period there are two possible stable asymptotic behaviors corresponding to the alternans, and also 2:1 rhythms. The computed values for the APD as a function of stimulation frequency are shown in figure 8.5d. The theoretical calculations show the appearance of alternans, but do not accurately give the APD for the range of stimulation frequencies for which alternans was experimentally observed. The theory also predicts a bistability between 1:1 and 2:1 dynamics and between 2:2 and 2:1 dynamics. Although bistability in conduction through the AV node was originally observed by Mines, we know of no other observations of this phenomenon in AV conduction, though it is observed during periodic stimulation of spontaneously beating aggregates of chick-heart cells.

To this point we have considered propagation of excitation in one dimension, in which the dynamics could be analyzed by consideration of a finite difference equation. As might be expected, physiologically important and mathematically more complex situations have also been studied. We briefly comment on two of these—the propagation of action potentials in nerve or other excitable systems, and the electrical activity in the small intestine.

A large body of work exists on the propagation of electrical activity in nerve using Hodgkin-Huxley or related models. Since nerve cells transmit excitation via nondecrementing action potentials, mathematical models of nerve cells formulated as partial differential equations should show stable propagating activity. However, to prove rigorously the existence and stability of traveling waves in concrete situations is a difficult problem. Some analytic results on the existence and stability of traveling wave solutions in one dimension have been obtained, but in realistic models of the cardiac conduction system only numerical simulations of conduction have been carried out, and analytic results are lacking.

As a final example, consider the electrical activity present in the small intestine. We have already mentioned slow-wave activity is present in the longitudinal muscle cells in the small intestine (figure 5.1). There is a gradient in the frequency of slow-wave intestinal activity, which, in humans, is about 12/min in the duodenum and decreases aborally to about 8/min in the terminal ileum. The coupling between the slow-wave oscillations leads to frequency plateaus as one moves down the intestine with phase shifts within a single plateau. The slow-wave oscillations are not associated with contraction unless there are superimposed spikes on the peak of the slow waves (chapter 5.1). In fasting individuals there are striking *migrating myoelectric complexes* (MMC) associated with intestinal contractions, which slowly traverse the length of the intestine in about 90 min (figure 8.6). As the MMC passes a given region of the intestine, regular bursts of activity arise at the frequency set by the slow-wave oscillation. This activity will last approximately 20 min out of the 90-min MMC cycle. At other times of the MMC cycle there are either no contractions or irregular contractions. Eating a meal abolishes

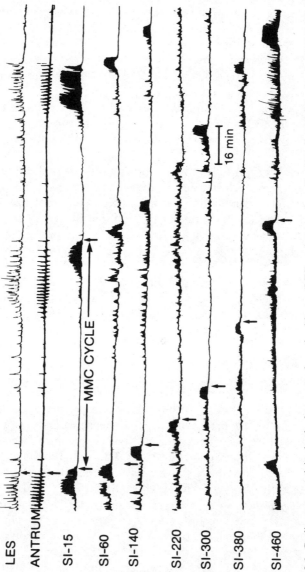

8.6. Cyclic motor activity in the lower esophageal sphincter (LES) and stomach, and migrating motor complexes (MMC) in the small intestine of dog recorded with strain gauges. Distances (in cm) indicated on small intestine strain gauges are from the pylorus. From Sarna (1985).

the MMC activity, and the two different types of intestinal contractions now occur. Peristaltic waves pass aborally down the small intestine for short distances (1–4 cm) at a velocity set by the apparent velocity of the slow waves. Segmentation contractions are localized contractions that are not coordinated with contractions above and below. The coordination of these various rhythmic activities is a complex task which is not yet fully understood.

8.2 Wave Propagation in a Ring of Tissue

Rings of tissue, cut from the jellyfish mantle, were studied early in this century by Mayer. Shortly thereafter, wave propagation in rings of cardiac tissue was studied by Mines and Garrey. In response to a single stimulus delivered to the tissue, waves of contraction were generated which spread in both directions from the point of stimulation and were annihilated when they collided. However, if stimuli were delivered at increasing stimulation frequency, a situation frequently arose in which a wave continued to propagate in one direction even after stimulation was discontinued. Calling the velocity of propagation v, the path length l, and the refractory time θ, a stable circulating wave is only possible provided $l/v > \theta$. The observation of such circulating or *circus waves* in cardiac tissue provided experimental evidence for the hypothesis that some cardiac arrhythmias may be due to circus or *reentry mechanisms*, in which excitation can repeatedly circulate through the cardiac musculature.

The mechanisms underlying the establishment of circulating excitation are not immediately obvious. One possibility is that there is *unidirectional block* in the cardiac tissue which restricts the conduction of activity to one direction but not the other.

Conduction block due to elevated refractory time can also lead to the establishment of circus excitation. Consider the situation in figure 8.7. Assume that the refractory time is θ at all points in a ring of tissue except for a small portion in which it is θ'. Stimulation delivered at a given point in the tissue propagates with a speed v, and the distance from the point of stimulation to the region of increased refractivity is l_1 and l_2, going clockwise and counterclockwise, respectively, where $l_2 > l_1$. Assume that two stimuli are delivered with a time t_s between them, where $\theta < t_s < \theta'$. The first stimulus passes through the tissue in clockwise and counterclockwise directions and annihilate. The second stimulus once again travels both clockwise and counterclockwise, but since $t_s < \theta'$, the clockwise excitation is blocked when it reaches the refractory region. However, provided $t_s > l_1/v + \theta' - l_2/v$, the second

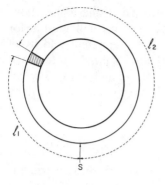

8.7. Schematic diagram for wave propagation in a ring of tissue. The refractory period is θ in the white region and θ' in the shaded region, where $\theta' > \theta$. If the tissue is stimulated two times from the point S with a time t_s between the stimuli, where $0 < t_s < \theta'$ and $t_s > l_1/v + \theta' - l_2/v$, then a counterclockwise circulating wave of excitation will be generated.

pulse will reach the region of increased refractoriness after it has become excitable, and a circulating wave of excitation will result.

In a tissue in which all points have the same refractory period, Wiener and Rosenblueth showed theoretically in 1946 that a circulating wave of excitation can be generated from two stimuli delivered to the tissue at different times and different positions, so that the second stimulus falls at the end of the refractory period of one of the spreading waves. As a consequence, the effect of the second stimulus is a spreading wave in only one direction. Now there are three simultaneous waves in the ring of tissue, two circulating in one direction and one in the other. Two of these waves moving in opposite directions will annihilate each other, leaving the third as a circulating wave of excitation. It is also easy to see that circulating waves will be possible in sheets of tissue with a barrier, provided the circumference of the barrier is sufficiently large, so $l/v > \theta$.

The recognition that circulating waves around a hole or barrier can be stably maintained is an important concept in cardiac physiology. Almost 70 years ago, Lewis showed that stable circulating atrial waves could exist around the openings of the great veins on the right auricle by mapping the cardiac action potentials on the surface of the heart of a dog. Lewis proposed that this circulating wave was associated with atrial flutter. Another example of circus movement occurs in patients with accessory pathways between the atria and ventricles. In these patients tachycardias sometimes arise in which circus movements are possible as a result of normal anterograde conduction through the AV node and retrograde conduction through the abnormal accessory pathway. A permanent cure for these tachycardias can be obtained by surgical section of the accessory pathway, thereby breaking the reentry loop. In other patients, tachycardias appear to arise from circus movement around an aneurysm or infarct.

8.3 Waves and Spirals in Two Dimensions

Consider a hypothetical experiment in which a circulating wave of excitation is set up in a ring of tissue; as the wave propagates the hole shrinks, eventually going to zero. What will happen to the wave? At first, it might seem that maintaining a circulating wave would be impossible since there is no way that the wave could continue to circulate around a hole of zero circumference. Yet, surprising as it may seem, waves can continue to circulate.

There are fundamental problems involved in the observation of waves in two and three dimensions. Whereas the propagation of a wave in one dimension can be monitored by comparatively few measuring devices along the length of the excitable system, in higher dimensions it is necessary to either observe the wave propagation visually or to have a large number of measuring devices to sample adequately the entire space. As we have already discussed, the study of wave propagation in excitable media has been facilitated by experiments on the Belousov-Zhabotinsky reaction, which displays target patterns and spirals (figure 1.12).

This complex chemical phenomenology may appear to be totally irrelevant for understanding spatial dynamics in physiology. However, we believe that understanding the propagation of these chemical waves will be central to our understanding of complex cardiac arrhythmias, and rhythms and arrhythmias in other spatially extended excitable tissues. In the rest of this section we give a summary of some of the main work on two-dimensional wave propagation in physiology.

An early description of circulating excitation in cardiac tissue in which there was no clear anatomical barrier was given in 1924 by Garrey, who studied circus movement in turtle hearts. He observed that "local faradization [electrical stimulation], confined to one spot on the auricle, started a circuit which coursed irregularly about that spot. The diameter of this intramuscular circuit was indefinite, something like 1 cm. or more. The entire remainder of the auricle responded to each circuit with a coordinate contraction repeated regularly 159 times a minute." This and other related observations formed the basis for the first theoretical study of reentry and fibrillation in 1946 by Wiener and Rosenblueth, who proposed a model in which space and time are continuous, and the state of the cardiac musculature was represented by a single variable. Following an instantaneous excitation, the tissue was refractory for a time, and then the tissue was excitable. Although they demonstrated circus movement around a sufficiently large barrier, their model did not appear to support circus movement in the absence of a barrier, but they could not prove this. Wiener and Rosenblueth obviously found Garrey's observations disturbing because they did not coincide with the theoretical model, and they raised the possibility that in Garrey's work "an artificial transitory obstacle was established unwittingly at the stimulated region."

Subsequent work has unequivocally established the presence of stable circus movement in two dimensions even in the absence of a barrier. The evidence comes from diverse experimental and theoretical work, which is summarized more fully in the Notes and References. Briefly, it revolves about the following facts.

1. Stable spiral waves are observed in nonliving excitable systems— e.g., in a two-dimensional iron-wire grid in nitric acid which was proposed by Nagumo and coworkers in 1963 as a physical model of nerve excitation, and in the Belousov-Zhabotinsky reaction.

2. Stable spiral waves are observed in living excitable systems. The first observation was made in 1965 by Gerisch, who observed spiral waves in aggregating slime molds. In the slime molds there is aggregation to a pacemaker which periodically releases a burst of cyclic adenosine monophosphate. In addition, in 1977 Allessie and coworkers demonstrated a circulating excitation in rabbit atrium, in the absence of any barriers using multiple electrodes to record the circus movement. Their *leading circle hypothesis* suggested that circus movement in the absence of a barrier could account for reentrant arrhythmias.

3. Spiral waves are observed in numerical simulations of cellular automata. In these systems, time and space are discrete. The state of each cell at a given time is a function of the states of neighboring cells at preceding times. In 1961 Farley and Clark first observed spirals in models of nerve networks. Moe and coworkers, trying to understand the mechanism of fibrillation, developed a cellular automata model of the myocardium—containing heterogeneous refractory times—that displayed complex circulating excitation in response to stimulation. This model was reexamined recently by J. M. Smith and R. J. Cohen, who studied its response to gradually increasing stimulation frequencies and observed complex periodicities as well as complex circulating excitation.

4. Spiral waves are observed in numerical simulations and theoretical analyses of continuous nonlinear systems.

In summary, in two-dimensional excitable media, simple target patterns as well as more complex spiral waves can be found. The evidence for this comes from experiments in chemical and biological systems and from numerical simulations and theoretical analysis of mathematical models of these systems. A detailed discussion of the physiological significance of this work as a model for fibrillation is left to section 8.5.

8.4 Organizing Cente   in Three Dimensions

Real biological systems are not one or two dimensional, but are really three dimensional. The three-dimensional nature of biological tissue should play an important role for tissues in which all three spatial dimensions are roughly comparable—for example, for the ventricles of the heart and possibly for cortical structures. The observation of excitation in three dimensions is difficult and the main results are principally theoretical. The main advances in understanding the geometry of propagation of excitation in three dimensions are due to Winfree and Strogatz.

One way to generate a three-dimensional wave is to translate the spiral waves perpendicular to the plane. The resulting *scroll waves* are shown in figure 8.8. In this case, the axis of the scroll wave extends to the boundary of the three-dimensional medium. The possibility also exists that the axis of the scroll can form a loop by joining the ends of

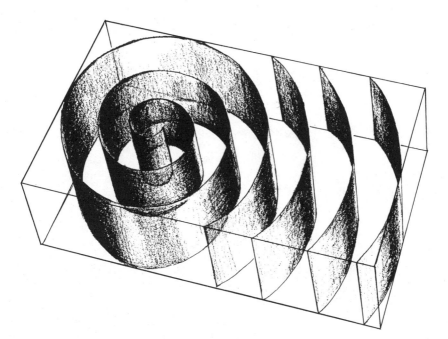

8.8. Three dimensional scroll wave. A thin section might appear as a spiral wave (see figure 1.12). From Winfree (1973a).

(a)

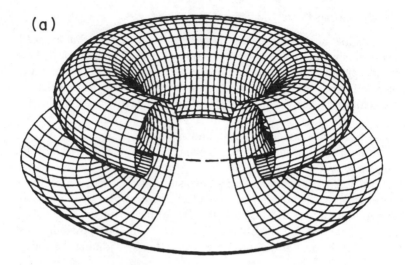

8.9. Three dimensional waves formed by joining the ends of the scroll wave. (a) Computer-drawn image. From Winfree and Strogatz (1984b). (b) (*right*) Belousov-Zhabotinsky reagent showing the same geometry. From Welsh, Gomatam, and Burgess (1983).

the axis, as illustrated in figure 8.9a. Figure 8.9b shows the Belousov-Zhabotinsky reaction in a three-dimensional medium. The resemblance to figure 8.9a is striking. The scroll ring of figure 8.9a, b is just the beginning of whole families of increasingly complex and linked rings that are theoretically possible.

8.5 Fibrillation and Other Disorders

Sudden cardiac death, in which the heart develops a fatal arrhythmia without warning, kills over 400,000 people yearly in the United States. In some instances this arrhythmia is a *bradycardia* (slow heart rate), but in the majority of cases it is a *tachycardia* (fast heart rate) that frequently originates in the ventricles. Although tachycardias are often initially rather regular, they generally degenerate into low-amplitude, irregular wave patterns on the electrocardiogram, reflecting the "fibrillation" present in the heart. Since the classic studies by Mines in 1914 and Wiggers and Wegria in 1940, it has been known that fibrillation can also be induced in the normal heart by electrical stimulation of the myocardium. Thus it is clear that fibrillation reflects abnormal organization of cardiac activity. The desire to understand this abnormal organization and to develop treatments to avert the establishment of fibrillation

(b)

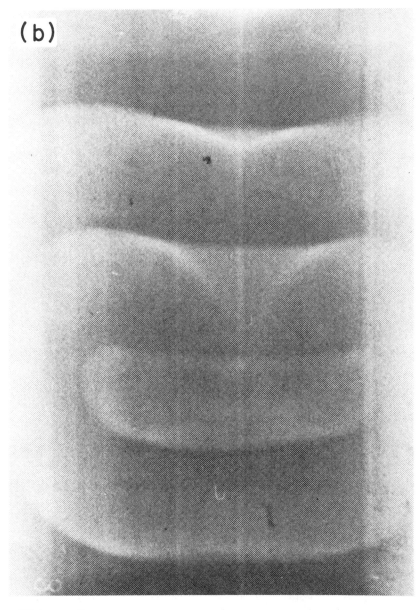

8.9 (*continued*)

underlies much of the research on the spatial organization of activity in excitable media discussed above.

An early description of fibrillation was given by Garrey in 1924. Since his description portends subsequent developments, we quote extensively:

> The general process of fibrillary contraction of heart muscle may be described as an incoordinate, disorderly and extremely bizarre contractile process in which normal systole and diastole no longer occur, the impression being given that individual fibers or groups of fibers are contracting independently (hence the name fibrillation). While certain regions of the fibrillating tissue are at rest, other adjacent area or areas widely separated from each other may show synchronous contraction. The surface of the fibrillating chamber shows areas of fine twitchings, of flickering and tremulous movements, combined with coarser undulating waves of muscular contraction which progress slowly through the muscle mass, moving now in one direction, now in another, being continually blocked in their progress by interference with other waves. In one and the same heart, every gradation may at times be seen, from the condition in which the coarse undulatory movements predominate, a condition to which Kronecker (1896) referred as a rolling movement which is comparable to a stormy peristalsis, to that in which the fibrillating tissue presents the appearance of a quivering mass or shows coarse, rapid fluttering movements with superimposed fine, fibrillary twitchings (Robinson, 1913). Riotous and chaotic as this fibrillation appears to be, analysis indicates that it may be aptly spoken of as a contractile *maelstrom*, for it appears that contractions are not independent of each other, but that the contractile impulse travels in a ringlike circuit repeatedly returning to and involving a given region after completion of each circuit (Garrey, Mines). This idea of a trapped wave, as presented by Garrey and by Mines, has been generally accepted by present day experimenters (c.f. infra).

Concerning a theoretical description of the mechanism that can lead to the fibrillation, Garrey is once again remarkably lucid:

> Impulses can spread in any and all directions, their progress being limited only by the preexistence or development of localized blocks within the tissue mass. Such blocks divert the impulse into other and more circuitous paths, and the area so blocked off can participate in contraction only when an impulse which has passed the other parts of the ventricle approaches it from another direc-

tion; this area in turn becomes a center from which the progress of contraction is continued, to be in turn diverted by other blocks. The existence of such blocks, and especially of blocks of transitory character and shifting location, has been noted in the experiments detailed above. These conditions make possible the propagation of the contraction wave in a series of ring-like circuits of shifting location and multiple complexity. It is in these *circus contractions*, determined by the presence of blocks, that we see the essential phenomena of fibrillation.

These observations are consistent with and anticipate much recent research on fibrillation. They are in accord with the suggestion by Moe and coworkers that fibrillation is due to the wanderings of multiple wavelets in the myocardium, and with the emphasis of the Russian school (as typified by the work of Krinskii) on the importance of "reverberators" and the fractionation of reverberators to give multiple reentry circuits. The recent emphasis by Winfree on circus motion as a basis for fibrillation is also consistent with Garrey's early discussions.

The development of microelectrodes and computers has made possible elegant physiological studies of tachycardia and fibrillation in the intact myocardium. Allessie and coworkers have observed circus motion in rabbit atria, and quite recently they have observed complex propagation that appears to be analogous to the multiple wavelets suggested from the computer studies of Moe. A dramatic example was obtained by Downar and colleagues by recording epicardial activity of the ventricles during cardiac surgery in humans. Rotating spiral waves, similar to those observed by Allessie in rabbit atria, were observed during an episode of ventricular tachycardia (figure 8.10). El-Sherif and coworkers have studied the patterns of activation for one to five days following coronary artery ligation in dogs. A pattern of excitation was observed in which "two circulating wavefronts advance in clockwise and cour rclockwise directions, respectively, around two zones (arcs) of functional conduction block." It is intriguing to speculate that these two circulating waves might be analogous to the clockwise and counterclockwise rotations that are often set up in the excitable Belousov-Zhabotinsky reaction (figure 1.12b).

From a practical standpoint, one of the crucial aspects of the study of fibrillation is to determine how it is established in cardiac tissue. One mechanism is by repeated periodic stimuli delivered to a single locus on the myocardium. Numerical simulations by Moe and coworkers and J. M. Smith and R. J. Cohen show that such periodic stimuli, if sufficiently rapid, will lead to dynamics resembling fibrillation following removal of the stimuli, in a medium with a variable refractory period.

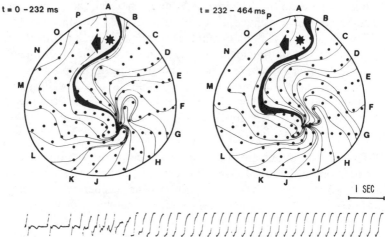

8.10. Epicardial mapping of ventricular activation recorded in humans during programmed stimulation. Lower panel shows three local epicardial electrograms in transition from programmed pacing (arrows) to sustained ventricular tachycardia. Upper panel
shows retrospective epicardial maps in which consecutive wave fronts of excitation, at
12-msec intervals, for two sequential beats of the tachycardia. The activation wheeled
around the apex, consistent with ventricular flutter. From Downar et al. (1984).

An intriguing result from the simulations is that complex periodicities
(such as alternans or other period-multiplied rhythms) result for stimulation frequencies slightly lower than those that gave rise to the fibrillation. These rhythms are due to the variable conduction around and
through regions of high refractory periods, and are not related to the
period-doubling cascades observed in nonmonotonic one-dimensional
maps.

Another result from these simulations shows that alternans is sometimes observed prior to fibrillation. This finding is of interest since it is
known from experiments in dogs that alternans in T-wave morphology
is associated with a lowered ventricular fibrillation threshold. Although
it is tempting to speculate that such alternans may be part of a period-
doubling cascade leading to chaos (i.e., fibrillation), a clear mechanism
that could generate such a cascade has not been proposed. As discussed
in section 8.1, the alternating response of cardiac tissue to stimuli at
high stimulation rates may be due to a period-doubling bifurcation

(which is not part of a cascade). Thus the appearance of alternans prior to fibrillation in response to rapid stimulation requires a careful and conservative analysis. At present the association of fibrillation with the chaos observed in one-dimensional maps is not justified.

A second mechanism for inducing fibrillation is to give a single electrical shock at a critical phase (called the *vulnerable phase* by Wiggers and Wegria) in the cardiac cycle during the refractory phase of the ventricles. This timing is clinically important because fibrillation occurring in the clinical context of frequent ectopy is often associated with the fall of ectopic beats during the repolarization of the ventricles, the *R on T phenomenon*. In addition, the induction of atrial fibrillation is often associated with a premature atrial contraction that falls during the repolarization phase of the atria. As Winfree has shown, a single stimulus falling during the refractory period could give rise to spiral-wave activity patterns. If the stimulus is delivered so that it generates a wave that intersects the trailing refractory zone of a propagated wave, the endpoints of the stimulated wave might roll up to give rise to two counterrotating spiral waves. Thus a single shock delivered to spontaneously oscillating homogeneous tissue might lead to a graded phase resetting, culminating in spiral waves. However, the phase-resetting argument presented by Winfree is not directly applicable to fibrillation since the usual medium for cardiac fibrillation is an excitable but not a spontaneously oscillating tissue. Nevertheless, the observation that circus movement can be set up in a homogeneous tissue is important and contrasts with other hypotheses based on heterogeneous refractory times.

In view of the complex nature of fibrillation, there has been interest in developing quantitative measures to characterize it. However, since fibrillation is not stationary, and fibrillation in different contexts may have different characteristics, quantitative studies are open to unavoidable problems. Computation of power spectra during ventricular fibrillation and autocorrelation during atrial fibrillation reveal strong periodicities, which may reflect the presence of regular periodic circus motion or localized pacemakers. It is important to recognize that autocorrelation functions, power spectra, or dimension calculations based on surface electrocardiograms are crude measures of the spatial organization of fibrillation.

The discussion to this point has dealt with fibrillation and circulating waves in cardiac tissue, but it is clear that any excitable tissue can support circus movement and fibrillation. Two tissues in which circus movement has been demonstrated are the rat cortex and the chicken retina. This work is based on earlier observations of waves of spreading

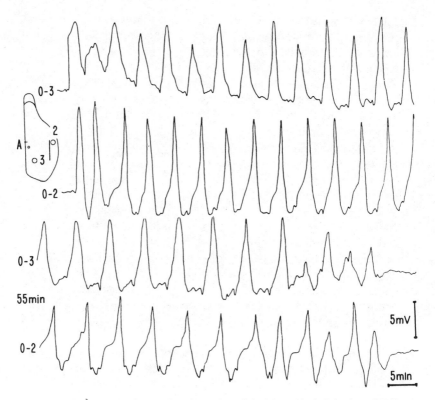

8.11. Circulating wave of spreading depression elicited by a single injection of KCl to the cortex of the rat. The circulating rhythm lasted for about 3 hr (55 min of the middle part of the record are omitted). During the last five complexes, there was KCl application by which the circulation of spreading depression was stopped. The small circle is the site of KCl injection, the large circles represent the sites of the recording electrodes, and the straight line represents an incision in the cortex. It was usually necessary to administer two KCl injections to initiate the circulating wave. From Shibata and Bures (1972).

cortical depression following application of potassium chloride to neural tissue. The slow spread of the waves (2–4 mm/min) is similar to the spread of waves in Jacksonian epilepsy and the scintillating scotoma of migraine headaches. Figure 8.11 shows recordings from two points on a rat cortex during passage of a circulating depression. The appearance of alternating activity levels is particularly intriguing. This may be related to the alternans described in section 8.1, and warrants further study.

Another organ in which circus movement might be expected is in the stomach, but we know of no documented description of circus movemen there. Recently, however, You and coworkers have written about a patient with gastrointestinal motility abnormalities (*tachygastria*), including severe nauseau, vomiting, epigastric bloating and pain, and weight loss. Recording of gastric activity from serosal electrodes indicated abnormally rapid rhythms, postulated to be caused by "the development of an abnormal ectopic or wandering pacemaker in the antrum." The symptoms were greatly improved by hemigastrectomy. However, histologic study of the antrum and a segment of the jejunum did not disclose any discernible lesions. A possible, but completely speculative mechanism for the malady would be fibrillation or circus movement, but much more detailed multielectrode recordings would be needed to confirm this speculation.

8.6 Summary

The spatial ordering of oscillations can differ from simple wavelike excitations emanating from a point source. In one dimension, if the stimulation frequency is too large, there will either be dropped beats (Wenckebach and other block phenomena) or there will be an alternation of response (alternans). In two dimensions, there can be propagating spirals having either one or several arms. In three dimensions, still more complex geometries called organizing centers are theoretically possible, but they are extremely difficult to observe experimentally. It is possible that complex dynamics in two and three dimensions in which there are multiple reentry circuits (spirals or organizing centers) underlie fibrillation.

Notes and References, Chapter 8

8.1 Wave Propagation in One Dimension

The cardiac rhythms in which there is partial blocking of propagation in the neighborhood of the AV node, resulting in grouped beating, are now called Wenckebach rhythms in honor of their discoverer (Wenckebach 1904). Early attempts at analysis of the mechanisms of Wenckebach rhythms were carried out by Mines (1913), Mobitz (1924), and Lewis and Master (1925), who recognized that the PR interval is often a decreasing function of the preceding RP interval. Mobitz (1924) and more recently others (Decherd and Ruskin 1946; Levy et al. 1974a,b· Glass, Guevara, and Shrier 1987; Shrier et al. 1987) have shown that the AV recovery curve can be used to compute the effects of periodic stimulation as a function of the frequency of the periodic stimuli. A similar theoretical model for AV heart block has been proposed by Keener (1981), who provided an understanding of the bifurcations of the finite difference

equations that model this arrhythmia (Keener 1980, 1981). For further discussions of Wenckebach rhythms, see Bellett (1971), Zipes (1979), and Marriott
and Conover (1983). Wenckebach-like rhythms are also observed in other
tissues such as the ureter (Weiss, Wagner, and Hoffman 1968).

Alternans rhythms, in which there is an alternation of action potential morphology, have also been observed in diverse tissues such as the ureter (Prosser,
C. E. Smith, and Melton 1955), canine antral circular muscle (Publicover
and Sanders 1986), and embryonic chick-heart cells (Guevara et al. 1984). These
experimental observations may be related to the electrical alternans phenomenon observed in electrocardiology (Bellett 1971). The theoretical analysis of
alternans using iterative techniques was pioneered by Nolasco and Dahlen
(1968), and our treatment follows along similar lines (Guevara et al. 1984). For
a discussion of the electrical restitution curve, see Boyett and Jewel (1978).

In this chapter we do not discuss in detail wave propagation in one-dimensional excitable media which originates from a point source in which there is
periodic propagation with the same period as the pacemaker. Such situations
have been extensively analyzed, notably in the papers of Rinzel and coworkers
(Rinzel 1980, 1981; Rinzel and Miller 1980; Rinzel and Maginu 1984). For other
important theoretical results on traveling wave solutions in one dimension, see
Carpenter (1979) and Feroe (1983). These references should be consulted for
details and for full citations to earlier work. In realistic models of the cardiac
conduction system, analytic results are not yet available, though numerical
simulations of conduction have been carried out (Joyner et al. 1983).

In other theoretical work specifically related to the intestine, the frequency
gradient of coupled nonlinear oscillators has been modeled by several groups,
and it has been possible to demonstrate the presence of frequency plateaus in
the mathematical models (Diamant, Rose, and Davison 1970; Sarna, Daniel,
and Kingma 1971; B. H. Brown et al. 1975; Patton and Linkens 1978; Sarna
1985). Recently, studies of a simplified mathematical model tor the small intestine have successfully obtained analytic results on the plateau phenomenon
(Kopell and Ermentrout 1983). Wave propagation underlying locomotion in
fishes (Kopell 1986) and the lamprey (Rand, A. H. Cohen, and Holmes 1988)
has also been modeled by coupled oscillators.

8.2 Wave Propagation in a Ring of Tissue

Rings of tissue from the jellyfish mantle were first studied early in this century by Mayer (1908). Shortly thereafter, wave propagation in rings of cardiac
tissue was studied by Mines (1913, 1914) and Garrey (1914, 1924).

The idea that conduction block due to elevated refractory time can also lead to
the establishment of circus excitation is due to Lewis (1920; see also Krinskii
1968; Marriott and Conover 1983).

Wiener and Rosenblueth (1946) have considered a model of tissue with a
homogeneous refractory time throughout, specifically within the context of a
tissue with a "hole" or nonconducting region. Lewis (1920, 1925) showed that
stable, circulating atrial waves could exist around the openings of the great
veins on the right auricle by mapping the cardiac action potentials on the

surface of the heart of a dog, and proposed that this circulating wave was associated with atrial flutter. Another example of circus movement is in Wolff-Parkinson-White patients in whom there is an accessory pathway between the atria and ventricles (Marriott and Conover 1983; Gallagher 1985). A permanent cure for tachycardias in these patients can be obtained by surgical section of the accessory pathway, thereby breaking the reentry loop (Wallace et al. 1974). Finally, some tachycardias appear to arise from circus movement around an aneurysm or infarct (Josephson et al. 1985).

8.3 Waves and Spirals in Two Dimensions

See Winfree (1980, 1987b) for a much more complete summary of the main work on two-dimensional wave propagation in physiology.

The observations by Garrey (1914, 1924), Mines (1913, 1914), and Lewis (1920, 1925) formed the basis for the first theoretical study of reentry and fibrillation by Wiener and Rosenblueth (1946). Selfridge (1948), working with the same model, was able to show that it did not support stable circus movement in the absence of an obstacle.

Stable spiral waves in nonliving excitable systems were apparently first observed by Nagumo, Suzuki, and Sato (1963). For a discussion of the results in this reference (which is difficult to obtain) and a full discussion of spiral waves in the Belousov-Zhabotinsky reaction, see Winfree (1980, 1987b). Additional observations on spiral waves in the Belousov-Zhabotinsky reaction are in Agladze and Krinsky (1982) and Muller, Plesser, and Hess (1985). Stable spiral waves are observed in aggregating cellular slime molds, as described by Gerisch (1965) and more recently by Durston (1973). The observation of circus movement in rabbit atria was carried out by Allessie, Bonke, and Schopman (1977).

Spiral waves have been observed in many studies of the dynamics of cellular automata (Farley and Clark 1961; Reshodko and Bures 1975; J. M. Greenberg, Hassard, and Hastings 1978; Madore and Freedman 1983). Studies of cellular automata as models of cardiac action potential propagation are found in Moe, Rheinboldt, and Abildskov (1964) and J. M. Smith and R. J. Cohen (1984).

Spiral waves are observed in numerical simulations and theoretical analyses of continuous nonlinear systems. Numerical results were obtained by Gulko and Petrov (1972) in a model of excitable biological systems, and by Winfree (1974) and Miura and Plant (1981) in a simplified model of excitable kinetics. Spirals were also observed in mathematical models of oscillating chemical reactions (Erneux and Herschkowitz-Kaufman 1975; D. S. Cohen, Neu, and Rosales 1978). A number of analytical results dealing with the existence and stability of spiral waves in two-dimensional media have been obtained (Kopell and Howard 1981; Hagan 1982; Mikhailov and Krinskii 1983; Keener 1986).

8.4 Organizing Centers in Three Dimensions

The original description of the scroll wave (Winfree 1973a) flowered 10 years later in the elaboration of a beautiful and detailed theory of spatial organization in three-dimensional excitable media (Winfree and Strogatz 1983a,b,c, 1984a,b), which is summarized and explained in a nontechnical way in Winfree

(1987b). It is still too soon to evaluate the practical importance of these theoretical results. However, observations of a simple three-dimensional geometry corresponding to theoretical expectations have been made in the Belousov-Zhabotinsky reaction (Welsh, Gomatam, and Burgess, 1983).

8.5 Fibrillation and Other Disorders

Mines (1914) and Garrey (1914, 1924) had deep insights into the origin of tachycardia and fibrillation. The earliest experimental work on fibrillation induced by electrical stimulation of the myocardium is by Mines (1914) and Wiggers and Wegria (1940). The role of premature atrial contractions in inducing atrial fibrillation is discussed in Bennett and Pentecost (1970), and the role of premature ventricular contractions in inducing ventricular fibrillation and sudden death (the R on T phenomenon) is discussed in Nikolic et al. (1982) and Hohnloser et al. (1984). A current perspective on sudden cardiac death is in Lown (1979). A suggestion by Winfree (1983b) that Mines accidentally killed himself in 1914 while performing phase resetting experiments on his own heart has received widespread circulation. Documentary evidence apparently inconsistent with this suggestion has been found and is in the Osler Library at McGill University (see also Winfree 1987b).

The hypothesis that fibrillation is due to the wanderings of multiple wavelets due to variable refractory times in the myocardium is due to Moe and coworkers (Moe and Abildskov 1959; Moe et al. 1964) and has recently been reexamined by J. M. Smith and R. J. Cohen (1984) and Allessie et al. (1985). Ritzenberg et al. (1984) have shown that the post stimulus fibrillatory-like responses in these models are due to the variable conduction around and through regions of high refractory period. The alternans in T-wave morphology associated with lowered ventricular-fibrillation threshold in dogs was shown by Adam et al. (1984). Similar ideas, but stressing the importance of "reverberators," have been popular in Russia (Krinskii 1968).

Direct mapping of the electrical activity during arrhythmia promises to clarify what is going on during tachycardia and fibrillation (Allessie et al. 1977; Allessie et al. 1985; Downar et al. 1984; El-Sherif 1985). Ideker and Shibata (1986) have recently used the mapping techniques to test hypotheses about the origin of tachycardia and fibrillation which have been proposed by Winfree (1983b, 1985, 1987b) and have found agreement with theoretical predictions. In particular, they have observed the establishment of both clockwise and counterclockwise rotating waves due to single shocks delivered during a critical phase of the cardiac cycle. The effects of a stimulus on an excitable, but nonoscillatory medium presents special problems for theoretical interpretation (Winfree 1987b).

The nonstationarity of fibrillation has been discussed by Wiggers (1940) and Goldberger et al. (1986). The computation of power spectra of ventricular fibrillation was carried out by Herbschleb et al. (1979) and Goldberger et al. (1986), and the autocorrelation of atrial fibrillation was reported by Battersby (1965).

One topic of contention is whether fibrillation is "chaotic" (Battersby 1965; J. M. Smith and R. J. Cohen 1984; Goldberger et al. 1986). Since there are no generally accepted operational definitions for fibrillation or chaos, this question could be debated endlessly.

Circus movement in the rat cortex was described by Shibata and Bures (1972) and in the chicken retina by Martins-Ferreira et al. (1974). The original observation of waves of spreading cortical depression following potassium chloride application was due to Leao (1944). A report of tachygastria is in You et al. (1981).

Chapter 9

Dynamical Diseases

We have proposed that diseases characterized by abnormal temporal organization be called *dynamical diseases*. In this chapter we briefly review the concept of dynamical disease. In section 9.1 we discuss the identification of dynamical disease. There is a large literature in which mathematical models for dynamical disease have been formulated, and this theoretical approach is discussed in section 9.2. An alternative approach, discussed in section 9.3, is to formulate biological models of dynamical disease and to analyze them theoretically. One eventual goal is to utilize principles and concepts from biological and mathematical models for diagnosis and for the rational design of therapies. This is discussed in section 9.4.

9.1 Identification of Dynamical Diseases

The normal individual displays a complex mosaic of rhythms in the various body systems. These rhythms rarely display absolute periodicity. Indeed, quantitative measurements of such rhythms as the heart rate and respiration frequently reveal much greater fluctuations in these systems than might be naively expected. Goldberger and colleagues have even suggested that the normal healthy dynamics are "chaotic" and disease is associated with periodic behavior. However, clear mechanisms which can lead to chaos (in the technical sense used in this book) for normal dynamics have yet to be enunciated. Whether or not one interprets normal dynamics as chaos or some other type of dynamic behavior, it is clear that many pathologies are readily identifiable by abnormal rhythmicities.

The signature of a dynamical disease is a marked change in the dynamics of some variable. Three types of qualitative changes in dynamics are possible and all have been observed: (1) variables that are constant or undergoing relatively small-amplitude "random" fluctuations can develop large-amplitude oscillations that may be more or less regular. Thus there may be the appearance of a regular oscillation in a physiological control system not normally characterized by rhythmic processes; (2) new periodicities can arise in an already periodic process;

and (3) rhythmic processes can disappear and be replaced by relatively constant dynamics or by aperiodic dynamics.

During a 30-year period starting in the late 1940s, Reimann studied periodic diseases in which the period was greater than 24 hours. In 1974 he wrote:

> Since 1947 I have collected reports of more than 2,000 examples of various medical disorders recurrent at weekly, fornightly, monthly or irregular intervals in otherwise healthy people. These repetitive disturbances are far more intense and disabling than those incident to the circadian rhythm. They are heritable and, with an exception, only one of several entities afflicts a family [Reimann 1963].

> Generally, and in current textbooks of Medicine, the proposed nosologic relation of the disparate entities is received skeptically and the subject is confused. I probably have not made my views clear, or my essays may not be read carefully or the idea is disregarded. Each entity is different clinically, but all have in common heredity, precise or irregular periodicity of short episodes of illness for decades, overlapping features, suppression of episodes during pregnancy, occasional replacement of one entity by another, occasional amyloidosis, similar serum complement defects and they resist therapy.

Reimann studied a host of different periodic diseases, including disorders that are associated with periodic recurrence of fever, periodic swelling of joints, periodic fluctuations of circulating blood cells, and periodic edema. Whether or not the disorders Reimann was considering share any common etiologic features, his emphasis on instilling order based on dynamic features of the disease was not popular. In a similarly frustrated tone, Crammer wrote in 1960: "Cyclic processes are facts of pathology as of physiology, and multiple rhythms occur in the single individual throughout the evolutionary tree. This knowledge does not seem to have penetrated clinical medicine in the way it has gained acceptance in the rest of biology, perhaps because so little is known of the underlying mechanisms of cyclic illnesses."

Our observations stress the importance of careful experimental documentation of the time-dependent behavior of physiological control systems in health and disease, particularly in response to changes in control parameters. Such observations not only provide important insight into the nature of the underlying control systems, but also place constraints on the features that proposed models must contain.

Unfortunately, published long time-series for physiological phenomena are uncommon, particularly in the recent clinical literature. In our research on dynamical diseases over the past decade, we have been struck by the great difficulty, if not impossibility, of identifying vital data sets containing long time-dynamics of numerous physiological variables in a single individual. Thus extended records of blood cell concentrations, hormone levels, blood pressure, labor rhythms, and so on are not easily found.

Although the accumulation of such long time-series of descriptive data is sometimes difficult in the current research environment, other reasons may exist for their absence. It is quite possible that both interesting and relevant dynamical changes are often observed but not published because their significance is not fully appreciated, or the dynamical changes are wrongly ascribed to environmental noise and/or experimental error. The pooling of data from different experiments or patients often obscures the presence of interesting dynamics in experimental and clinical time records. A fundamental property of chaotic systems is that their dynamics are exquisitely sensitive to small changes in either the values of the control parameters and/or initial conditions. In view of the extensive range of biological variability, it is not surprising that even at the best of times the observed dynamics between two experiments or patients are not precisely the same. By pooling time-series, one could easily submerge interesting dynamics into a monotonous and humdrum noisy sea.

A wealth of dynamical behavior ranging from periodic to irregular, noiselike oscillations can be readily observed both experimentally and clinically in physiological control systems. Although many of these situations are familiar to the physiologist, the universal and fundamental aspects of their rich dynamical fabric does not yet appear to be fully appreciated. The importance of these qualities becomes more evident when it is realized that relatively simple nonlinear mathematical models have these same properties, thus implying that dynamic complexity may be the norm rather than the exception in living systems. Significant and striking dynamical features of normal and pathological physiological systems exist, and a continuing study of these in the future is likely to be extremely fruitful in its yield of insight into fundamental aspects of physiological control.

9.2 Formulation of Mathematical Models
for Dynamical Diseases

The notion that it is possible to formulate mathematical models that capture qualitative characteristics of human disease is certainly not new. Prior to the coining of the term "dynamical disease," numerous studies existed in which mathematical models for physiological systems had been proposed, and altered dynamics in the mathematical model as a result of bifurcations (not always characterized as such) were identified. For example, studies of cardiac conduction were undertaken by Mobitz and by van der Pol and van der Mark in the 1920s and Wiener and Rosenblueth in the 1940s. There were extensive studies of oscillatory instabilities in negative feedback systems in the 1960s.

Throughout this book we have given a number of examples from studies characterizing bifurcations in mathematical models of physiological systems. Currently, those familiar with the basics of nonlinear dynamics should be able to propose mathematical models for physiological systems which display dynamic properties such as limit cycle oscillation or "chaos." What is not so easy, however, is to make sure that the origin of the normal or pathological oscillation is associated with a hypothesized physiological mechanism.

The formulation of mathematical models for physiological system is a powerful tool for fixing ideas and developing insight into physiological dynamics. However, such modeling in the absence of concrete applications to clinical or experimental systems will simply remain interesting hypotheses. What is needed is the continual interplay between theory and experiment which has characterized the physical sciences. Although some may feel that biology is different from the physical sciences, and will never be amenable to the type of sharp theoretical analysis common there, we disagree. We are convinced that the rich dynamical phenomena in physiology and medicine can be understood using techniques being developed to understand dynamics in nonlinear systems.

9.3 Development of Biological Models for Dynamical Diseases

The great advantage in developing a biological model for a dynamical disease is that it is possible to undertake systematic manipulations that are simply not possible to undertake in human beings. Examples of such biological models are the insertion of a length of tubing into the carotid artery to increase the delay time between oxygenation of blood in the lungs and the arrival of the blood in the brain stem (as a model for Cheyne-Stokes respiration), the cycling of white blood cells in grey collie dogs (as a model for the periodic fluctuations of white cells in cyclical neutropenia), and the periodic stimulation of Purkinje fibers in vitro (as a model for cardiac-arrhythmia generation).

The potential pitfall in biological models for dynamical diseases is that the observed altered dynamics may mimic those observed clinically, but for different reasons. Since the possible types of bifurcation from stable or oscillating states are often limited, it may be possible to establish qualitatively similar dynamics in many different ways. Despite this, combined systematic experimental and theoretical work on biological models of dynamical disease is essential. Even if the particular model is shown to be inappropriate as a model for a particular disease, the careful working-out of the theory for bifurcations in dynamics as a function of system parameters is bound to be of interest. Unfortunately, the current structuring of most institutions of higher learning, combined with the inadequate training of both theoreticians and experimentalists and the current nature of research support, conspire to make the requisite interdisciplinary studies logistically difficult to design, implement, and complete

9.4 Diagnosis and Therapy

A goal for many of those interested in the application of nonlinear dynamics to physiology is to develop practical techniques for diagnosing pathological conditions and deciding on rational therapeutic strategies for treating them. The biological aspects of our own research has largely centered on the short- and long-range control of the cell cycle and cell proliferation with applications to hematological disorders, the dynamics of simple neural feedback systems, and the effects of single and periodic perturbations on physiological oscillations. In the course of this work, a number of potential applications of nonlinear dynamics to clinical medicine have occurred to us. As illustrations of these, we briefly describe several possible approaches drawn from our own work and that of others.

In many clinical situations, periodic stimuli are delivered to the patient as a therapeutic measure. Examples are drug administration and the use of electronic cardiac pacemakers and mechanical ventilators. In these circumstances, establishing a well-controlled, stable relationship between the imposed rhythm and the bodily rhythms can be difficult to achieve.

For example, in some diabetic patients it can be difficult to establish appropriate schedules for insulin administration. In these patients, periodic insulin administration combined with regular eating and exercise schedules is ineffective in maintaining blood glucose within normal limits. Rather, there can be apparently irregular fluctuations (for example, in blood glucose monitored upon arising). In such patients it will be necessary to develop protocols for insulin administration based on a knowledge of ambient blood-sugar levels and an understanding of the dynamics of the glucose-control system.

Implementation of sensing mechanisms to aid in the effective therapeutic use of mechanical ventilators and cardiac pacemakers has been carried out. These sensing mechanism allow feedback from the patient to the controlling device to facilitate operation and to avoid dangerous competition between imposed and intrinsic rhythms. However, a detailed understanding of the dynamics of these devices as they are used clinically is an extremely difficult theoretical problem due to this bidirectional coupling.

A novel suggestion for using periodic stimulation clinically has been made by a colleague, John Milton at the Montreal Neurological Institute. He suggests that it may be possible to suppress tremor by periodic stimulation. Such an effect would be analogous to suppression of oscillation in cardiac cells by periodic depolarizing stimuli, as demonstrated by Guevara in 1984. Similarly, properly timed doses of medication might be able to suppress seizures in regularly cycling epileptics.

Another clinically important situation involving interactions between endogeneous and external rhythms involves circadian rhythms. The observation that circadian rhythms are often altered in patients with affective disorders has led to attempts to treat these patients by restoring the normal phase relationship between the intrinsic sleep wake and the normal 24-hour cycle. Phase shifting of the circadian clock can be accomplished by light, by small changes in sleep pattern imposed over several days, and by drugs. It is important to recognize that changes in the circadian rhythm may be one effect rather than the cause of the affective disorder, so that treatment of the circadian abnormality will not necessarily cure the affective disorder. Finally, since it has been demonstrated that benzodiazepines can affect the circadian rhythms in hamsters, the possibility has been raised that the effects of jet lag can be reduced by appropriate administration of drugs, thereby phase resetting the circadian rhythm. However, setting appropriate doses and schedules for drug administration is not yet possible.

Studies of the cell cycle and blood cell control have led to a number of proposals concerning patients with blood diseases. Clinical data and the study of the dynamics of feedback systems for the regulation of hematopoesis (see section 4.6) have led to the recognition that oscillations and perhaps chaotic dynamics may be expected in patients with blood disorders. In a study of periodic hematopoiesis (PH, also known in the early literature as cyclical neutropenia), Mackey in 1978 was able to relate the dynamic changes found in the blood-cell populations of PH patients to a bifurcation in the dynamics of the pluripotential stem cell (PPSC) population, accurately predicting the period of the oscillations observed as well as a number of other characteristics of the disease. The bifurcation was hypothesized to arise from an (abnormal) cellular death during the proliferative phase in the PPSC population. The hypothesis that cell death during proliferation in the PPSC can give rise to oscillations was subsequently confirmed using ^{89}Sr to induce cell death in W/W^v mice. A subfraction of aplastic anemia patients is thought to suffer from the same pathology (abnormal cell death), and the hypothesis for the origin of PH suggests that some of these aplastic-anemia patients should show oscillations in their stem-cell numbers and more mature hematopoietic cells during their recovery. This phenomenon has been observed by Morley in unpublished work.

The recognition that patients with blood diseases may frequently cycle has important practical implications. In the absence of any therapy, a subpopulation of CML patients display prominent cycling of their white blood-cell levels between normal and elevated values (see figure 1.8 and section 4.6). An unpublished study of patient records from the Oncology Day Centre of the Royal Victoria Hospital, Montreal, suggests that many CML patients being treated with chemotherapy were also "cyclers," and thus their white cell counts might have spontaneously decreased to normal even without the chemotherapy. The possibility that patients with CML may be cycling regularly or chaotically confounds treatment. In fact, populations of patients with CML show the same survival statistics now as they did during the period from 1910 to 1948 in spite of the advent of sophisticated chemo- and radiotherapies. One possible explanation (see section 4.6) is that some patients die because of the therapy much sooner than if they had been left alone, while others have their lifespan increased! This hypothesis presupposes a white blood-cell control system operating in a chaotic domain, and would suggest that therapy will become effective only when the intricacies of this control system have been truly mastered.

Insight from nonlinear dynamics for therapy need not always arise from situations in which periodic or chaotic oscillations occur. For example, in patients suffering from marrow hypoplasia of various origins—for example, from radiotherapy or chemotherapy—one might think that the more the demand on the PPSC to produce differentiated cells was decreased, the more advantageous to the recovery of the PPSC and their marrow descendents it would be. Pursuing this original line of thought, the dynamic response of a continuous maturation proliferation model for the erythroid-production control system has been treated as a nonlinear optimal control problem. A central point of interest was the hematopoietic response after a massive decrease in the number of proliferating PPSC, as found in many of these cases of marrow hypoplasia, including aplastic anemia. Based on this analysis, it was possible to show that there is an optimal level of the peripheral demand for mature erythrocytes that will allow the maximum possible repopulation of the PPSC population. Surprisingly, the optimal demand level on the PPSC for efficient repopulation is not zero. This insight was used to design a simple therapy for the treatment of these pathologies, based on decreasing the demand for differentiated erythroid precursor cells. This can be accomplished either by having the patients breathe in an oxygen-enriched environment or by administering a massive transfusion of erythrocytes. These procedures have been successfully tested at the Academy of Medicine, Krakow, Poland, but large clinical trials have not been undertaken.

9.5 Summary

Many human diseases are characterized by unusual and complex dynamic behavior. The analysis of the mechanisms underlying such diseases must inevitably deal with a theoretical analysis of the observed dynamics. Approaches to study these problems involve the formulation of theoretical and biological models of the disease. A long-term goal for researchers is to help provide novel diagnostic and therapeutic strategies for the treatment of humans. It is our hope that this book will help to stimulate research in these areas.

Notes and References, Chapter 9

9.1 Identification of Dynamical Diseases
Normal variability of the heart rate is described in Kitney and Rompelman (1980), Akselrod et al. (1981), Kobayashi and Musha (1982), and normal variability of respiration is described in Goodman (1964). The hypothesis that this normal fluctuation of the heart rate is associated with "chaos" is due to Goldberger and colleagues (Goldberger et al. 1985, 1986; West and Goldberger 1987; Goldberger and Rigney 1988). A mechanism which can give rise to highly complex and perhaps chaotic dynamics in normal physiological control systems is the interactions between multiple feedback loops controlling a single variable (Glass, Beuter, and Larocque 1988). In recent work, available only in preprint form at the time of writing of this book, Goldberger and Rigney (1988) summarize their position as follows: "cardiac arrest represents a bifurcation from the fractal, chaotic dynamics of the normal heartbeat to the pathologic periodicities of the dying heart." It is important to recognize, however, that numerous clinical observations going back 40 years identify apparently irregular rhythms, which were even called "chaotic" by Katz (1946), as precursors to sudden cardiac death (Nikolic, Bishop, and Singh 1982; Hohnloser et al. 1984). Many other pathologic rhythms in the human heart, such as atrial fibril-

lation, multifocal atrial tachycardia, and frequent ventricular ectopy frequently appear to be highly irregular (Bellett 1971; Marriott and Conover 1983; Goldberger and Goldberger 1986).

Early discussion of periodicities in disease has been given by Reimann (1963, 1974), Crammer (1959, 1960) and C. P. Richter (1965). We first used the term "dynamical disease" in 1977 (Mackey and Glass 1977), and have returned to this theme often (Glass and Mackey 1979a; Mackey and an der Heiden 1982; Guevara et al. 1983; Mackey and Milton 1987; an der Heiden and Mackey 1988).

9.2 Formulation of Mathematical Models
for Dynamical Diseases

Formulation of mathematical models for disease has been attempted many times in the past. The study of mathematical models for cardiac arrhythmias was pioneered by Mobitz (1924), van der Pol and van der Mark (1928) and Wiener and Rosenblueth (1946). Models for the control of blood cell production were proposed by King-Smith and Morley (1970) and Lasota (1977). Studies of instabilities in negative feedback systems were undertaken by Grodins (1963), Milhorn (1966), and Stark (1968). Throughout this book we have presented theoretical models for a variety of pathological conditions, including Cheyne-Stokes respiration (section 4.5), chronic myelogenous leukemia (section 4.6), and AV heart block (section 8.1). In general, we anticipate that an understanding of the changes in dynamical behavior associated with disease will necessarily entail a formulation of suitable mathematical models.

9.3 Development of Biological Models
for Dynamical Diseases

For a biological model of Cheyne-Stokes respiration, see Guyton Crowell, and Moore (1956). For a discussion of the cycling of white blood cells in grey collie dogs, see J. E. Lund, Padgett, and Oh (1967), and Dale, Alling, and Wolf (1972). Jalife and Moe (1976, 1979) developed biological models for the cardiac arrhythmia parasystole.

9.4 Diagnosis and Therapy

An experimental demonstration of the suppression of periodic activity in heart cells by periodic stimulation has been given by Guevara (1984) and Guevara, Shrier, and Glass (1988). Phase resetting of the circadian clock in humans by light has been reported by Czeisler et al. (1986). Turek and Losee-Olsen (1986) induced phase resetting of the circadian clock in hamsters using benzodiazepines. The possibility for treating affective illness by manipulating the circadian clock is discussed in several papers in the collection by Wehr and Goodwin (1983). Light therapy in various contexts is being pursued actively by Lewy and coworkers (Lewy, Sack, and Singer 1985). A discussion of the relevance of phase resetting of circadian clocks to jet-lag is in Winfree (1986).

A model for periodic hematopoesis was proposed by Mackey (1978), and observation of cycling of blood cells following induced cell death using radioactive strontium was given in Gurney, Simmons, and Gaston (1981). Statistics on the survival of patients with chronic myelogenous leukemia are in Shimkin, Mettier, and Bierman (1950) and Wintrobe (1976). A model for survival with this disease is in Lasota and Mackey (1980).

Formulation of the continuous maturation proliferation model of the cell cycle was carried out by Mackey and Dormer (1981, 1982), and application of this model to aplastic anemia was carried out by Lasota, Mackey, and Wazewska-Czyzewska (1981). Therapies suggested by this study have been implemented by Wazewska-Czyzewska (1984).

Somehow, a myth has arisen (which we believe is accepted by the great majority of practicing biologists) that detailed mathematical and theoretical analyses are not appropriate in biology. Certainly the mathematical training of most biologists and physicians is minimal. Yet if the complex dynamic phenomena that occur in the human body were to arise in some inanimate physical system—let us say in a laser, or liquid helium, or a semiconductor—they would be subjected to the most sophisticated experimental and theoretical study.

We have had two aims in writing this book. This first is to make physical scientists aware of the enormous complexity and beauty of dynamic phenomena in physiology and medicine. The second is to show physiologists and physicians that the techniques of nonlinear mathematics are applicable, and in some cases essential, to the analysis of dynamic phenomena in physiology.

Mathematical Appendix

In this book we have emphasized theoretical concepts and biological problems rather than given details about computational techniques. Yet, underlying the general theory is a large body of mathematical methods subsumed under the rubric "Theory of Dynamical Systems." This area is generally considered as an advanced area in mathematics and as such it is not normally taught at the undergraduate level. Accordingly, texts in this area may not be suitable for many readers of this book.

The use and understanding of the principles we have discussed will be enormously facilitated if the reader attempts to develop some technical abilities that will enable him or her to formulate and analyze finite difference equations and differential equations as models of particular biological systems. The mathematical concepts needed to carry through such analyses do not require a sophisticated mathematical background, and we have in fact taught undergraduate biology and physiology students the basics for many years. In this Appendix we sketch out briefly the central computational techniques and give some problems. We discuss separately differential equations and finite difference equations.

A.1 Differential Equations

Mathematical models in the physical and biological sciences are often formulated as ordinary differential equations of the form

$$\frac{dx_i}{dt} = f_i(x), \qquad x_i(t = 0) = x_i(0), \qquad i = 1, 2, \ldots N, \qquad \text{(A.1)}$$

where $x_i(t)$ represents the ith variable and the function $f_i(x)$ gives the time evolution of $x_i(t)$. In the simplest situation the functions on the right-hand side of equation (A.1) are *linear*, that is, all the variables on the right-hand side appear only to the first power. In this situation equation (A.1) can be written

$$\frac{dx}{dt} = Ax, \qquad x(t = 0) = x_0, \qquad \text{(A.2)}$$

where A is an $N \times N$ matrix, x represents an N-vector, and x_0 a vector of initial conditions.

The solutions of linear ordinary differential equations are completely understood and can be readily calculated. To do this one solves the *characteristic equation*, which is given by

$$\det(A - pI) = 0, \tag{A.3}$$

where I is the identity matrix and "det" means determinant. The *characteristic equation* is an Nth order polynomial in p, and in general it will have N different roots, called *eigenvalues*, with non-zero real parts. In this situation the solutions of equation (A.2) usually can be written

$$x_i(t) = \sum_{i=1}^{N} c_i e^{p_i t}. \tag{A.4}$$

where p_i represents a root of the characteristic equation (A.3) and c_i represents a constant, which may be complex and is normally set from the intial conditions.

The case of exponential decay considered in chapter 2 is an example of equation (A.2) when there is only one variable. When there are two variables, equation (A.2) can be written as

$$\frac{dx_1}{dt} = ax_1 + bx_2, \tag{A.5}$$

$$\frac{dx_2}{dt} = cx_1 + dx_2,$$

where a, b, c, d represent constants. By differentiating in equation (A.5) and then substituting, equation (A.5) can be rewritten in the form

$$\frac{d^2x_1}{dt^2} - (a + d)\frac{dx_1}{dt} + (ad - bc)x_1 = 0. \tag{A.6}$$

Equation (A.6) is a linear, second-order ordinary differential equation since the highest order of the derivative is 2 and all derivatives and powers appear as linear terms. In second- (or higher) order linear differential equations it is convenient to determine the characteristic equation by substituting $x(t) = Ce^{pt}$ and solving the resulting characteristic equation.

Example 1. Radioactive materials decay according to the equation

$$\frac{dx}{dt} = -\alpha x,$$

where α is a constant. The half-life of radioactive tritium is 4.5×10^3 days. Give the formula for the amount of radioactive tritium as a function of time if N is present at $t = 0$.

Solution. Radioactive tritium decays exponentially so that $x(t) = Ne^{-\alpha t}$. The value of α can be calculated by determining the time $t_{1/2}$ that it takes until $x(t) = 0.5\ N$. Thus $\alpha = \ln 2/t_{1/2}$, or in this case $\alpha = 1.5 \times 10^4\ \text{days}^{-1}$.

Example 2. An intravenous administration of a drug can be described by a two-compartment model with compartment 1 representing the blood plasma and compartment 2 representing body tissue (figure A.1). This system can be modeled by the differential equations,

$$\frac{dx_1}{dt} = -(k_1 + k_2)x_1 + k_3x_2,$$

$$\frac{dx_2}{dt} = k_1x_1 - k_3x_2,$$

where x_1 and x_2 are the concentrations of the drug in compartments 1 and 2, respectively, and k_1, k_2, and k_3 are positive constants that give the flow between compartments. Solve this equation for x_1 as a function of time, starting from an initial condition $x_1(0) = C$, $x_2(0) = 0$ for the special case $k_1 = 0.5$, $k_2 = k_3 = 1$.

Solution. The characteristic equation is

$$\begin{vmatrix} -(k_1 + k_2) - p & k_3 \\ k_1 & -k_3 - p \end{vmatrix} = 0,$$

which can be solved to give

$$p_{1,2} = \frac{-(k_1 + k_2 + k_3) \pm [(k_1 + k_2 + k_3)^2 - 4k_2k_3]^{1/2}}{2}.$$

Substituting the given values of k_1, k_2, and k_3, we find $p_1 = -1/2$, $p_2 = -2$ so that

$$x_1(t) = Ae^{-t/2} + Be^{-2t}.$$

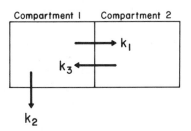

Compartment 1   Compartment 2

k_1

k_3

k_2

A.1. A compartmental model for transport of drugs between the blood plasma (compartment 1) and body tissue (compartment 2). The k_i represent rate constants for transport.

From the initial conditions, we find that

$$A + B = C.$$

$$\frac{dx_1}{dt} = -3C/2 = -A/2 - 2B.$$

This can be solved to determine $A = C/3$ and $B = 2C/3$, so that the solution of the problem is

$$x_1(t) = \frac{C}{3} e^{-t/2} + \frac{2C}{3} e^{-2t}.$$

Example 3. A damped pendulum, in the limit of small-amplitude oscillations, is described by the differential equations,

$$\frac{d^2\theta}{dt^2} + k \frac{d\theta}{dt} + \omega^2\theta = 0,$$

where θ is the angular displacement from the vertical, k is a positive constant proportional to the friction, and ω is the angular frequency. The angular frequency in turn is $(l/g)^{1/2}$, where l is the pendulum length and g is the acceleration due to gravity. Assume that $4\omega^2 > k^2$. At $t = 0$, $\theta = 5°$ and $\frac{d\theta}{dt} = 0$. What is θ as a function of time?

Solution. The characteristic equation is

$$p^2 + kp + \omega^2 = 0,$$

which has two roots,

$$p_{1,2} = \frac{-k \pm (k^2 - 4\omega^2)^{1/2}}{2}.$$

In this case the two roots are complex conjugates with a negative real part. Using the identity $e^{ia} = \cos a + i \sin a$, where $i = \sqrt{-1}$, the boundary conditions can be used to set constants in the same manner as in example 2. The solution is an exponentially decaying oscillation given by

$$\theta(t) = 5 \exp\left(\frac{-kt}{2}\right)\left[\cos\frac{\alpha t}{2} + \frac{5k}{\alpha} \sin\frac{\alpha t}{2}\right],$$

where $\alpha = (4\omega^2 - k^2)^{1/2}$.

The analysis of the linear ordinary differential equations provides a basis for the stability analysis of *nonlinear ordinary differential equations* in the neighborhood of *steady states* (also known as *equilibrium points* or *fixed points*), which are defined as points for which the derivatives of all variables are equal to zero. At a steady state, x^*, of equation (A.1),

$$\frac{dx_i^*}{dt} = f_i(x^*) = 0, \qquad i = 1, 2, \ldots, N, \qquad (A.7)$$

In the neighborhood of the steady state, the dynamics are given by

$$\frac{dx}{dt} = A(x - x^*), \qquad (A.8)$$

where the elements of the matrix A are defined as

$$a_{ij} = \left.\frac{\partial f_i}{\partial x_j}\right|_{x^*}. \qquad (A.9)$$

The eigenvalues of the matrix A can once again be computed using equation (A.3). These eigenvalues are useful in characterizing the qualitative dynamics in the neighborhood of the steady state. If the real parts of all the eigenvalues are less than zero, the steady state is *asymptotically stable*, and it is asymptotically approached in the limit $t \to \infty$ from all initial conditions in the neighborhood of the steady state. If the real part of one or more of the eigenvalues is positive, then the steady state is *unstable*. If the largest real part of the eigenvalue(s) is zero, the steady state is called *neutrally stable*.

Let us consider the dynamics in the neighborhood of steady states in two dimensions. If the origin of the axes is translated to the steady state, then the linearized equations in the neighborhood of the critical point is given by equation (A.5), and the eigenvalues are computed to be

$$p_{1,2} = \frac{a + d \pm [(a - d)^2 + 4bc]^{1/2}}{2}. \qquad (A.10)$$

The geometry of flows in the neighborhood of the critical point depend on the eigenvalues. The various generic cases have been named as follows:

Case 1. *Focus*, p_1 and p_2 are complex conjugates with nonvanishing real parts

$$a + d \neq 0, \qquad (a - d)^2 < -4bc.$$

Case 2. *Node*, p_1 and p_2 are real with the same sign,

$$|a + d| > [(a - d)^2 + 4bc]^{1/2}, \qquad (a - d)^2 > -4bc.$$

Case 3. *Saddle point*, p_1 and p_2 are real with the opposite signs,

$$|a + d| < [(a - d)^2 + 4bc]^{1/2}, \qquad (a - d)^2 > -4bc.$$

NODE

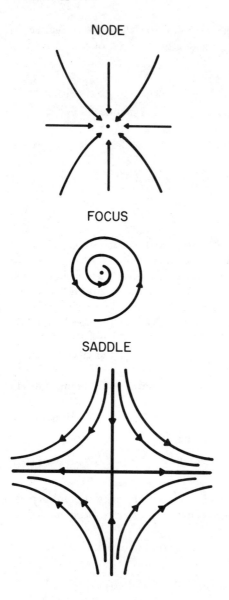

FOCUS

SADDLE

A.2. The three main types of steady states for two-dimensional ordinary differential equations. Stable nodes and foci are shown. For unstable nodes and foci, the trajectories are directed outward, away from the steady state.

Nodes and foci can be either stable or unstable. Saddle points are always unstable.

As discussed in chapter 2, it is common to plot the evolution of the system as time proceeds by sketching the trajectories in the phase space for different initial conditions. In figure A.2 we show the phase plane portraits of foci, nodes, and saddle points.

Bifurcations are associated with changes in the number and/or stability of steady states or other limit sets. For example, the *Hopf bifurcation* is associated with two complex eigenvalues crossing the imaginary axis. Determination of whether this corresponds to a supercritical or subcritical bifurcation (chapter 5) can (in principle) be carried out algebraically, though the computations may become horrendous. A second simple bifurcation is associated with the splitting of a single stable steady state into three steady states—a saddle point and two stable steady states.

The number and types of steady states are restricted by the geometry of the phase space. An important topological result due to Poincaré places restrictions on the steady states in two-dimensional vector fields. In practical situations, the dynamics are restricted to a finite connected region of phase space, and the trajectories on the boundary of the region are directed into the region. In this situation, in two dimensions, if we call $\mathcal{N}$, $\mathcal{F}$, and $\mathcal{S}$ the numbers of nodes, foci, and saddle points, respectively, then by the Poincaré Index Theorem

$$\mathcal{N} + \mathcal{F} - \mathcal{S} = 1. \tag{A.11}$$

Extensions of this result to higher-dimensional phase spaces and phase spaces of different topology are possible.

Example 4. A system in which there is mutual inhibition (see chapter 4) can be given by the differential equations,

$$\frac{dx}{dt} = \frac{\theta^n}{\theta^n + y^n} - x,$$

$$\frac{dy}{dt} = \frac{\theta^n}{\theta^n + x^n} - y.$$

If $\theta = 1/2$, there is a steady state at $x^* = y^* = 1/2$. Discuss the bifurcations and sketch the flows in the phase plane as n varies.

Solution. The characteristic equation at the steady state is readily calculated from equation (A.9) is

$$\begin{vmatrix} -1 - p & -\dfrac{n}{2} \\ -\dfrac{n}{2} & -1 - p \end{vmatrix} = 0.$$

The eigenvalues of this equation are $p_{1,2} = -1 \pm n/2$. The steady state is a stable node for $n \leqslant 2$ and a saddle point for $n > 2$. The trajectories in the two-dimensional phase space can be sketched (figure A.3). At

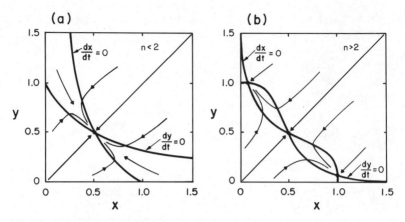

A.3. Schematic representation of the phase plane for mutual inhibition. (a) For $n < 2$ there is a single stable steady state. (b) For $n > 2$, there are three steady states, a saddle point and two stable nodes. Which node is reached in the limit $t \to \infty$ depends on the initial condition.

$n = 2$ there is a bifurcation in which the single steady state splits into a saddle point plus two stable steady states.

Example 5. The van der Pol oscillator is given by the equations

$$\frac{dx}{dt} = \frac{1}{\varepsilon}\left(y - \frac{x^3}{3} + x\right),$$

$$\frac{dy}{dt} = -\varepsilon x, \qquad \varepsilon > 0.$$

Describe the bifurcations as a function of ε, and sketch the phase plane for the case $0 < \varepsilon \ll 1$.

Solution. There is a single steady state at $x = y = 0$. The characteristic equation is

$$\begin{vmatrix} \dfrac{1}{\varepsilon} - p & \dfrac{1}{\varepsilon} \\ -\varepsilon & -p \end{vmatrix} = 0.$$

from which the eigenvalues can be readily computed to give

$$p_{1,2} = \frac{1}{2\varepsilon} \pm \frac{1}{2}\left(\frac{1}{\varepsilon^2} - 4\right)^{1/2}.$$

Therefore, for $\varepsilon > 1/2$ there is an unstable focus, and for $0 < \varepsilon < 1/2$ there is an unstable node. For $0 < \varepsilon \ll 1$, there is a limit cycle as shown in figure A.4.

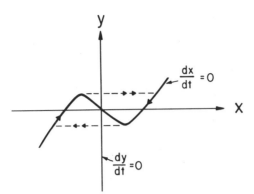

A.4. Schematic diagram of the phase plane for the van der Pol equation with $\varepsilon \gg 0$. A stable limit cycle is set up. The jump from one branch of the cubic to the other along the dotted lines is rapid.

The examples and calculations have concentrated on the analysis of the stability and bifurcations of steady states. Unfortunately, except in one-dimensional ordinary differential equations, knowledge about the number and stability of steady states is not sufficient to give a complete description of the global topological organization of the dynamics. Indeed, proofs of comparatively simple topological properties of dynamics, such as the uniqueness and stability of limit cycle oscillations in two and more dimensions are often very hard to find, and there is a great reliance on numerical methods to study nonlinear dynamics.

In contrast to ordinary differential equations, in which the right-hand side is a function of the current values of the variables, in delay differential equations the right-hand side may be a function of the value of variables at some time in the past. In physiological systems with feedback, time delays often arise because of the time needed to transmit information from receptors to effector organs (see sections 4.5 and 4.6). The analysis of the dynamics in time-delay equations poses many theoretical problems of great current interest. Here we concentrate on the comparatively simple problem of local stability analysis.

Consider the delay differential equation,

$$\frac{dx}{dt} = f(x, x_\tau), \tag{A.12}$$

where $x_\tau = x(t - \tau)$. Note that for this, in contrast to ordinary differential equations, initial conditions must be specified as initial functions,

$$x(t') = f(t') \quad \text{for} \quad t' \in [-\tau, 0],$$

and delay differential equations, in spite of their apparent simplicity, are actually infinite dimensional systems.

As usual, the steady states x^* of equation (A.12) are defined implicitly by $f(x^*, x^*) = 0$. Thus, using a Taylor series expansion of equation (A.12) in the neighborhood of a steady state x^*, and considering only first order terms, we obtain

$$\frac{dz}{dt} = Az + Bz_\tau, \tag{A.13}$$

where

$$z = x - x^*, \qquad A = \frac{\partial f}{\partial x}\bigg|_{x^*}, \qquad B = \frac{\partial f}{\partial x_\tau}\bigg|_{x^*}. \tag{A.14}$$

To examine the local stability of equation (A.12) in the neighborhood of a steady state x^* is equivalent to examining the solutions of equation (A.13) for the local stability of $z = 0$. Thus we make the *ansatz* $z = e^{\lambda t}$ which, with equation (A.13), gives

$$\lambda = A + Be^{-\lambda \tau}. \tag{A.15}$$

Generally, the eigenvalue λ is complex, $\lambda = \mu \pm i\omega$, and in order for local stability of $z = 0$ to be assured, we must have $\mu < 0$.

In 1950 Hayes gave a complete treatment for the conditions under which $\mu = Re(\lambda) < 0$, and these may be summarized by the following:

$$|A| > |B|$$

or

$$|A| < |B| \qquad and \qquad \tau < \frac{\cos^{-1}\left(-\dfrac{A}{B}\right)}{(B^2 - A^2)^{1/2}}, \tag{A.16}$$

where the principal value $[0 \leqslant \cos^{-1}(-A/B) \leqslant \pi]$ is taken. This condition defines the situation under which a steady state x^* of equation (A.12) will be locally stable. Though Hayes' treatment is rather complicated, a partial understanding for these criteria can be obtained from the following considerations.

We go back to equation (A.15) and consider the case when λ is purely imaginary, that is, $\lambda = i\omega$, so $\mu = 0$. This is, then, just the condition for neutral stability. Substituting $\lambda = i\omega$ into equation (A.15) gives

$$i\omega = A + Be^{-i\omega \tau}, \tag{A.17}$$

which is equivalent to

$$i\omega = (A + B \cos \omega \tau) - iB \sin \omega \tau. \tag{A.18}$$

Equating real and imaginary parts of equation (A.18) gives two equations,

$$-A = B \cos \omega\tau, \tag{A.19a}$$

and

$$\omega = -B \sin \omega\tau, \tag{A.19b}$$

which are to be solved for ω and τ. Squaring both equations and adding gives

$$\omega = (B^2 - A^2)^{1/2}. \tag{A.20}$$

Furthermore, from equation (A.19a) $\omega\tau = \cos^{-1}(-A/B)$, so

$$\tau = \frac{\cos^{-1}\left(-\dfrac{A}{B}\right)}{(B^2 - A^2)^{1/2}} \tag{A.21}$$

defines exactly the value of the time delay τ at which $\mu \equiv 0$, given values of A and B. Comparison of equations (A.20) and (A.21) with the Hayes criteria in equation (A.16) shows the connection. Often, but not always, as a parameter in equation (A.12) is varied so that a steady state x^* loses its stability [the conditions in equation (A.16) are violated], a Hopf bifurcation takes place with a pair of complex conjugate eigenvalues crossing from the left-hand to the right-hand complex plane. The inequalities in equation (A.16) can be obtained only from a finer analysis.

Equations (A.20) and (A.21) offer some interesting insights into the period of oscillation when stability is lost. At precisely the set of parameters (A, B, τ) defined by equation (A.21), there is a solution $z(t) = e^{i\omega t}$ of equation (A.13), where the angular frequency ω is given by equation (A.20). Since angular frequency ω and period T are related by $\omega = 2\pi/T$, equation (A.21) can be written in the alternative form,

$$T = \frac{2\pi\tau}{\cos^{-1}\left(-\dfrac{A}{B}\right)}. \tag{A.22}$$

From this it is readily derived that if A and B are the same sign,

$$2\tau \leqslant T \leqslant 4\tau, \tag{A.23a}$$

whereas, if A and B are different signs,

$$2\tau \leqslant T. \tag{A.23b}$$

Therefore, in the time-delay equation in equation (A.12), it is possible to place bounds on the period of the oscillation at the point of instability.

Example 7. The time-delay differential equation

$$\frac{dx}{dt} = \frac{\theta^n}{\theta^n + x_\tau^n} - x \qquad x > 0$$

represents a negative feedback system in which a substance, x_τ, decays exponentially but is produced by a monotonically decreasing feedback function that depends on the value of x at a time τ in the past. Determine the stability criteria when $\theta = 1/2$. For $n \gg 1$, what are the stability criteria and what is the period of the oscillation at the instability?

Solution. There is a steady state at $x^* = 1/2$. Doing a power-series expansion around this point and defining $z = x - 1/2$, we have, to first order,

$$\frac{dz}{dt} = -\frac{n}{2} z_\tau - z.$$

Applying Hayes's criteria [equation (A.16)], we find that the steady state is stable if

$$2 > n,$$

or

$$n > 2 \qquad \text{and} \qquad \tau < \frac{\cos^{-1}\left(-\dfrac{2}{n}\right)}{\left(\dfrac{n^2}{4} - 1\right)^{1/2}}.$$

For large values of n, the solution will be stable provided $n\tau < \pi$. At the point of instability, the period is 4τ. This shows that steady states in negative feedback systems with time delays are destabilized by increasing the gain of the feedback (here n) or the time delay, τ. Note that we have not provided analytic evidence that this bifurcation is a super-critical Hopf bifurcation, but further analytic studies show this to be the case.

A.2 Finite Difference Equations

One approach to analyzing differential equations in higher dimensions is to consider the properties of maps that represent the return of a cross section of flow to itself. Such maps are conveniently written as finite difference equations,

$$x_i(t + 1) = f_i(x(t)), \qquad i = 1, 2, \ldots, N, \qquad \text{(A.24)}$$

where $x_i(t)$ represents the value of the ith component at a time t, and f_i is a nonlinear function. An analysis of the stability of steady states in finite difference equations follows along the same lines as in differential equations. Assume that there is a steady state x^*, which is defined by the relationship

$$x(t + 1) = x(t) = x^*. \qquad \text{(A.25)}$$

Then equation (A.24) can be linearized in the neighborhood of the steady state to obtain

$$x(t + 1) = A(x(t) - x^*), \qquad \text{(A.26)}$$

where A is an N × N matrix and the elements of A are given as

$$a_{ij} = \left. \frac{\partial f_i}{\partial x_j} \right|_{x^*}. \qquad \text{(A.27)}$$

The eigenvalues of A are once again found by solving the characteristic equation,

$$\det(A - pI) = 0. \qquad \text{(A.28)}$$

The steady state, x^*, is stable if all the eigenvalues lie inside the unit circle. (A complex number $a + ib$ lies inside the unit circle if $a^2 + b^2 < 1$). A Hopf bifurcation occurs if two complex conjugate eigenvalues simultaneously cross the unit circle.

In one and two dimensions, finite difference equations display much richer dynamics than ordinary differential equations. This is because the trajectories of the time evolution of differential equations must be continuous, and this precludes the presence of cycles or chaos in one-dimensional differential equations and chaos in two-dimensional differential equations. However, both cycles and chaos can be found in one-dimensional finite difference equations. As we have discussed, one dimensional finite difference equations arise naturally as mathematical models in biology and the natural sciences. Moreover, comparatively simple one-dimensional finite difference equations yield enormously rich bifurcations under parameter changes, and in some cases a detailed mathematical analysis of the bifurcations has been possible. In the remainder of this Appendix we discuss exclusively the dynamics in one-dimensional finite difference equations. For notational convenience, we indicate subsequent iterates by subscripts.

As a simple example, we consider the linear finite difference equation

$$x_{t+1} = ax_t. \tag{A.29}$$

By iterating this equation we find

$$x_{t+2} = ax_{t+1} = a(ax_t) = a^2 x_t,$$

$$\dots \tag{A.30}$$

$$x_{t+n} = ax_{t+n-1} = a^n x_t.$$

Thus, if $|a| < 1$, then the iterates x_t will approach 0 in the limit $t \to \infty$. If however, $|a| > 1$, then the values x_t will be unbounded in the limit $t \to \infty$.

In one-dimensional finite difference equations, it is frequently convenient to iterate the equation graphically. This graphical iteration can be used to calculate the time evolution even in circumstances when the algebraic computation of the iterates is not feasible. In figure A.5 we show the iteration of equation (A.29) for the situation in which $0 < a < 1$, and also the situation in which $a > 1$. The geometrical decrease and increase are evident.

Another way of thinking about equation (A.29) is to think of it as resulting from a linearization about the steady state $x_{t+1} = x_t = 0$. This steady state is stable for $|a| < 1$ and unstable for $|a| > 1$, in accord with the stability criteria discussed earlier for the general case in N dimensions.

In the general one-dimensional finite difference equation

$$x_{t+1} = f(x_t), \tag{A.31}$$

a steady state, $x_{t+1} = x_t = x^*$, will be stable if $|(\partial f/\partial x)_{x^*}| < 1$, and unstable if $|(\partial f/\partial x)_{x^*}| > 1$. The reason for this is clear and relates to the linear expansion of the function at the steady state or, alternatively, to the value of the eigenvalue of the linearized equation at the steady state.

Starting from an initial condition x_0, one can iterate equation (A.31) to generate a sequence $x_0, x_1 = f(x_0), \dots, x_m = f^m(x_0)$. A periodic orbit of period n will arise if $x_{t+n}^* = x_t^*$, $x_{t+j}^* \neq x_t^*$ for $1 \leqslant j < n$. Let

$$p = \frac{d}{dx}\bigg|_{x=x_t^*} f^n(x) = \prod_{i=1}^{n} \left(\frac{\partial f}{\partial x}\right)\bigg|_{x=x_t^*}. \tag{A.32}$$

The stability of the periodic orbit is determined by the value of $|p|$; for $|p| < 1$ the orbit is stable, and for $|p| > 1$ it is unstable. When $|p| = 1$

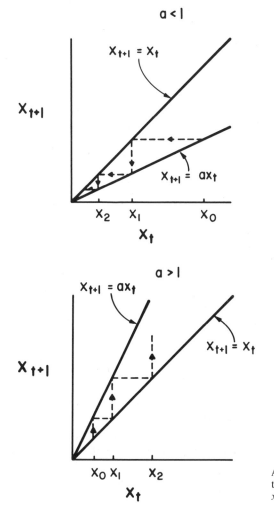

A.5. Graphical iteration of the finite difference equation $x_{t+1} = ax_t$.

there is a bifurcation of the periodic orbit. Generically, there are two definite possible bifurcations depending on the sign of p. Suppose the map f depends on a parameter μ, and that at $\mu = \mu_0$, $|p| = 1$. When $p = 1$, for values of μ close to μ_0 there is either no periodic orbit (say for $\mu > \mu_0$), or there are two orbits of period n, one stable and the other unstable (say for $\mu < \mu_0$). In this case we say that there is a *tangent bifurcation* at $\mu = \mu_0$. This means that as μ passes through its critical value, a new stable periodic orbit, as well as an unstable periodic orbit, appears.

When $p = -1$, depending on the sign of $\mu - \mu_0$ for μ small, there is either a stable orbit of period n or a stable orbit of period $2n$ and an unstable orbit of period n. The point $\mu = \mu_0$, where $p = -1$, is a *period-doubling bifurcation* point. This means that as μ passes through its critical value, the period of a stable and hence observable oscillation doubles.

Example 8. For the equation

$$x_{t+1} = ax_t(1 - x_t) \qquad 0 \leqslant a < 4, \qquad 0 \leqslant x_t \leqslant 1, \qquad \text{(A.33)}$$

determine all steady states, the range of values of a for which each steady state is stable, and the type of bifurcation that occurs when the steady state loses stability.

Solution. By setting $x_{t+1} = x_t$ and solving the resulting quadratic equation, we find that there is a steady state at $x^* = 0$ for all values of a, and at $x^* = \dfrac{a-1}{a}$ for $a \geqslant 1$. The slope at the steady state $x^* = 0$ is a, from which we see that $x^* = 0$ is a stable steady state for $0 \leqslant a < 1$ and an unstable steady state for $a > 1$. At $a = 1$ there is a bifurcation, but it is not a tangent bifurcation even though the slope at the steady state is 1, since there are two steady states (one stable and the other not, if one considers the steady state at $x^* < 0$) both for $a < 1$ and $a > 1$. For the steady state at $x^* = (a - 1)/a$, the slope at the steady state is $2 - a$. Thus this steady state is stable for $1 \leqslant a < 3$ and unstable for $a > 3$. At $a = 3$ the slope is equal to -1 and there is a period-doubling bifurcation with a stable period 2 cycle arising as a increases through the value $a = 3$.

Example 9. For the equation

$$x_{t+1} = x_t + b \sin 2\pi x_t \qquad \text{(mod 1)}, \qquad b > 0,$$

determine all steady states, determine the range of values of b for which each steady state is stable, and determine the type of bifurcation that occurs when the steady state loses stability.

Solution. The steady states are easily seen to be at $x^* = 0$ and $x^* = 1/2$. The slope at the steady state at $x^* = 0$ is $1 + 2\pi b$, which is always unstable for $b > 0$. The slope at the steady state at $x^* = 1/2$ is $1 - 2\pi b$. This steady state loses stability via a period doubling bifurcation at $b = 1/\pi$.

One of the remarkable features of finite difference equations is that they sometimes display complex sequences of bifurcations that depend on the general geometric features of the functions on the right-hand side of the equations. Thus different classes of equations that exhibit well-defined geometric features have been identified and the bifurcations analyzed. We now consider the bifurcation sequences that are found in two classes of equations: (1) single-humped functions defined on an interval, and (2) *circle maps*, which map the points on the circumference of a circle into itself.

In chapter 2 we have already discussed some of the properties for single-humped functions such as the quadratic map (A.33). As the parameter a increases, there are cascades of period-doubling bifurcations. The value of a at the first and second period-doubling bifurcations can be calculated analytically. The sequence of period-doubled orbits 1,2,4,8, ... converges by the value $a = 3.57$ ... and for values of a greater than this, new periodic orbits not in this sequence can be found. However, the sequence of periodic orbits as a continues to increase is well understood and was called the *U-sequence* (U for universal) by Metropolis, Stein, and Stein in 1973. Up to orbits of period 6, the U-sequence is 1,2,4,6,5,3,6,5,6,4,6,5,6. As the period of the orbit increases, the number of windows of a for which this orbit can be found also increases, giving rise to infinite numbers of windows of unimaginably (to a biologist) small size in which stable periodic orbits can be found. In addition, there exist values of a for which "chaotic" dynamics can be proven, given some technical definition of chaos. From a practical point of view, the very narrow windows at which stable periodic cycles appear preclude the observation of high periodicities in all but the most controlled experiments in which it is possible to minimize noise. In experimental work in physics, some observations of successive period doubling and the U-sequence have been made in diverse systems such as hydrodynamic instabilities and chemical oscillations. In biology, the observation of such behavior has been made in diverse mathematical models (see chapter 4, for example), but experimental observations of period doublings and chaos are more limited (see chapter 7).

In addition to maps of the interval with a single extremum, there is a well-developed theoretical understanding of the global organization of bifurcations in finite difference equations that map the points on the circumference of the circle into itself. Such functions arise naturally in biological systems—for example, during a periodic stimulation of a biological oscillator (see chapter 7). A circle map is of the form

$$\phi_{t+1} = f(\phi_t) \quad (\text{mod } 1), \tag{A.34}$$

where ϕ_t is a point on the unit circle S^1, and f is a function that may be nonlinear. If a circle map is continuous, it can be characterized by a number called the topological degree, which represents the number of times ϕ_{t+1} goes around the unit circle as ϕ_t goes around it once. The significance of the topological degree is that in periodic forcing of strongly attracting limit cycle oscillations, the dynamics can frequently be described by either circle maps of topological degree 1 or 0 (Section 7.4). The dynamics of equation (A.34) can be partially characterized by a rotation number. In equation (A.34) calling

$$\Delta\phi_t = f(\phi_t) - \phi_t, \tag{A.35}$$

the rotation number is

$$\rho = \lim_{n \to \infty} \sup \frac{1}{N} \sum_{t=1}^{N} \Delta\phi_t. \tag{A.36}$$

If there is a periodic solution of the finite difference equation, the rotation number is rational.

A conceptual understanding of some of the properties of the rotation number can be obtained by a consideration of differential equations defined on a torus (figure A.6). Provided there are no fixed points of the flow, the dynamics can be analyzed by consideration of a cross section to the flow. The map that takes points on this cross section back to itself is called the *Poincaré map*. Since the cross section is topologically a circle, the Poincaré map is a circle map. Moreover, since trajectories cannot cross, the Poincaré map is a one-to-one, invertible map. The rotation number gives the average rotation in the ϕ coordinate for one rotation in the θ coordinate (see figure A.6). The rotation number for any initial condition on the cross section must be the same. Periodic forcing of nonlinear oscillations with low-amplitude stimuli frequently can be described by invertible circle maps. In this case, the Arnold tongue structure shown in figure 7.5 is found. However, as the stimulation amplitude increases, this simple Arnold tongue structure is destroyed, and there are exceedingly complex bifurcations that are still not completely understood.

To illustrate what happens, we consider a model equation

$$\phi_{t+1} = \phi_t + \tau + b \sin(2\pi\phi_t), \tag{A.37}$$

where b and τ are constant. Equation (A.37) is continuous for all values of b. However, at the value of $b = 1/2\pi$, the map becomes nonmonotonic. Thus for $b > 1/2\pi$ the map is no longer a one-to-one invertible map of the circle, though it is still of topological degree 1.

A plot of the phase-locking zones as a function of b and τ is shown in figure A.7. For $b < 1/2\pi$, the Arnold tongue structure of figure 7.5 is

(a)

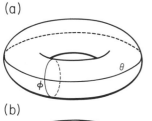

(b)

A.6. (a) A torus. The location of any point on the surface can be specified by two coordinates, ϕ and θ. (b) A trajectory on the surface of the torus. A cross section to the flow is shown. The Poincaré map gives ϕ_{t+1} as a function of ϕ_t. The rotation number counts the number of rotations in the ϕ coordinate for each rotation in the θ coordinate.

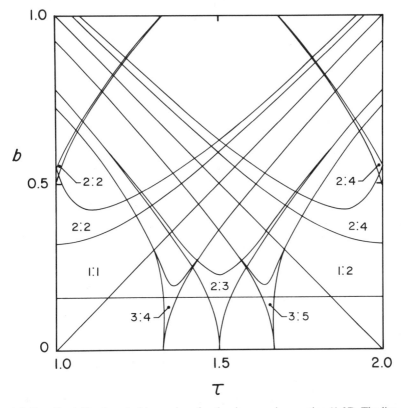

A.7. Locally stable phase-locking regions for the sine map in equation (A.37). The line at $b = 1/2\pi$ separates the region in which the map is a one–one invertible map ($b < 1/2\pi$) and noninvertible ($b > 1/2\pi$). The widths of some of the regions is so small as b increases that the boundaries are collapsed into a single line. There is bistability where two zones overlap. In the nonlabeled regions are phase locked, quasiperiodic, and chaotic dynamics. From Glass et al. (1983).

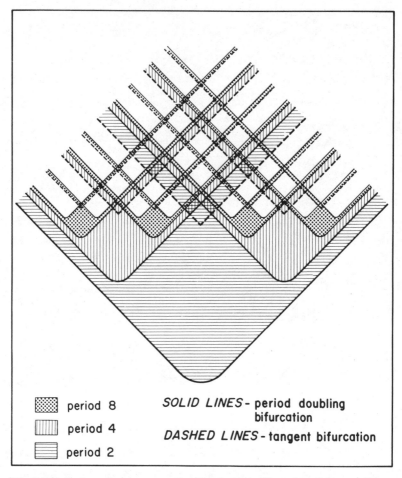

period 8

period 4

period 2

SOLID LINES - period doubling
bifurcation

DASHED LINES - tangent bifurcation

A.8. Schematic figure for the zones of period-doubling bifurcations in maps with two quadratic extremal points. This scheme has been found in cubic maps and circle maps. From Bélair and Glass (1985).

maintained. For the region $b > 1/2\pi$ in which the map is noninvertible, the structure is very different. Each Arnold tongue present for $b < 1/2\pi$ extends into the region $b > 1/2\pi$, splitting into two branches.

Consequently, extensions of Arnold tongues can cross, leading to a situation in which two different periodic orbits associated with different rotation numbers are found for the same values of the parameters. The other main feature is that there are complex sequences of bifurcations that are present in the Y-shaped regions formed by the extensions of each Arnold tongue. Each of the Arnold tongues has a complex geometrical arrangement of period-doubling bifurcations displaying self-similarity, which is schematically represented in figure A.8. Beyond the accumulation points of the period-doubling sequences, there are chaotic dynamics. The same topological structure depicted in figure A.8 was conjectured to be present in each of the Arnold tongues by Glass and Perez in 1982. In some sense this structure is an unfolding of bifurcations in maps with one parameter and one extremum to maps with two parameters and two extrema. This example is of interest because it displays a correspondence with periodically forced chick-heart cells (chapter 7).

The analysis of bifurcations of circle maps of topological degree different from 1 has not yet been carried out in great detail. However, the primary motivation for examination of this problem to date has come from the potential applicability to biological systems where maps of topological degree 0 arise naturally as a consequence of phase-resetting properties of biological oscillators that have been experimentally observed (chapter 6). The same sequence of period doublings shown in figure A.8 are also observed in degree 0 maps.

Since most results concerning the global organization of bifurcations have been found using a combination of numerical and advanced topological techniques, problems amenable to simple analytic solutions are not easy to find. The following two problems illustrate some features of the theory.

Example 10. Consider the circle map of topological degree 0,

$$\phi_{t+1} = b \sin 2\pi\phi_t \quad \text{(mod 1)}.$$

For what value of b is there a cycle of period 2 that passes through the two extrema?

Solution. The situation is shown in figure A.9. The extrema are at $\phi_t = 1/4$ and $\phi_t = 3/4$. Assuming $\phi_0 = 1/4$, we have $3/4 = b \sin \pi/2$, from which we find immediately that $b = 3/4$.

Example 11. Consider the circle map,

$$\phi_{t+1} = \phi_t - b \sin 2\pi\phi_t + \tau \quad \text{(mod 1)}.$$

Calculate the boundaries of the region in which there is a stable periodic orbit of period 1 and rotation number 1, and characterize the bifurcations on these boundaries.

Solution. Say there is a fixed point at ϕ^*. Then, for a period 1 orbit with rotation number equal to 1, we must have

$$1 + \phi^* = \phi^* - b \sin(2\pi\phi^*) + \tau.$$

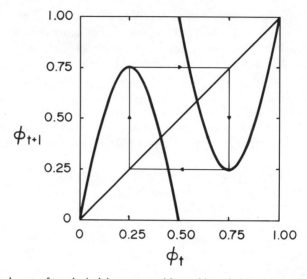

A.9. A circle map of topological degree zero with a stable period-2 orbit passing through
the two extrema (see example 10).

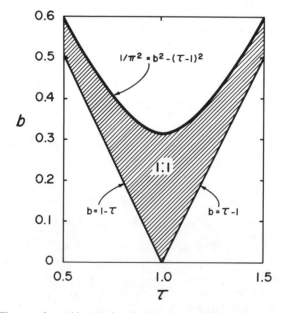

A.10. The zone for stable 1:1 phase locking for the circle map in example 11.

Provided $\left|\dfrac{\tau-1}{b}\right| \leqslant 1$, there will be a solution of this equation which is

$2\pi\phi^* = \sin^{-1}\left(\dfrac{\tau-1}{b}\right)$. The locus of the tangent bifurcation is found

by setting $\left.\dfrac{\partial\phi_{t+1}}{\partial\phi_t}\right|_{\phi^*} = 1$. From this we compute $2\pi b \cos 2\pi\phi = 0$, and

as b increases, keeping τ fixed, there is a tangent bifurcation along the lines $b = 1 - \tau$ and $b = \tau - 1$. As b continues to increase, the period 1 solution loses stability by a period-doubling bifurcation. This occurs

when $\left.\dfrac{\partial\phi_{t+1}}{\partial\phi_t}\right|_{\phi^*} = -1$. When this occurs, we compute that $-1 = 1 -$

$2\pi b \cos 2\pi\phi^*$. Substituting in the value for ϕ^*, we find that a period-

doubling bifurcation occurs along the hyperbola $\dfrac{1}{\pi^2} = b^2 - (\tau - 1)^2$.

The region in which there is a stable period 1 solution with rotation number 1 is shown in figure A.10.

A.3 Problems

Most of these problems are derived from published papers about oscillation and chaos in biological systems. In the interest of encouraging readers to undertake calculations on their own, we do not give the sources of the equations here (though most can be found with just a bit of digging). The problems are of different levels of difficulty and some are quite hard. Most (but not all) of them have been successfully solved by our undergraduate physiology students. Students who have access to computers will benefit from numerically simulating the dynamics.

1. In the differential equation,

$$\frac{dx}{dt} = \sin x - \alpha x \qquad x \geqslant 0, \alpha \geqslant 0,$$

discuss the bifurcations as a function of α. Starting from any initial condition, describe the dynamics as $t \to \infty$.

2. The differential equation,

$$\frac{d\phi}{dt} = \Omega - A \sin \phi,$$

where ϕ is taken modulo 2π and Ω and A are positive constants, has been considered as a model for two coupled, spontaneously oscillating neurons. ϕ is the phase difference between the activity in the two neurons. Discuss the qualitative dynamics and bifurcations as a function of Ω and A.

3. The "Brusselator," which has been proposed as a model of biochemical oscillations, is described by the differential equations,

$$\frac{dx}{dt} = a - bx + x^2y - x,$$

$$\frac{dy}{dt} = bx - x^2y,$$

where x and y are positive variables and a and b are positive constants. Determine the steady state and describe the stability as a function of a and b. What type of bifurcation occurs when the steady state loses stability?

4. The equations

$$\frac{dx}{dt} = 1 - xy^\gamma,$$

$$\frac{dy}{dt} = 4xy^\gamma - 4y,$$

where x and y are positive variables and γ is a positive constant, have been proposed as a model of glycolytic oscillations. Solve for the steady state, determine its stability, and classify it (node, focus, or saddle point) as a function of γ.

5.   (A) Based on the eigenvalues at the steady state, classify the different steady states in three dimensions, and sketch the trajectories in the neighborhood of each.

   (B) Suppose that a differential equation is defined in a ball in three dimensions and that the trajectories on the boundary of the ball are directed into it. There is a single steady state. Which ones, out of the steady states found in (A), may it be?

6. The following has been proposed as a model for feedback inhibition:

$$\frac{dx_1}{dt} = \frac{\theta^m}{\theta^m + x_N^m} - x_1,$$

$$\frac{dx_i}{dt} = x_{i-1} - x_i, \qquad i = 2, 3, \ldots, N.$$

Solve for the steady state and determine the criteria for a Hopf bifurcation as a function of N and m, when $\theta = 1/2$.

7. The differential equation,

$$\frac{dx_i}{dt} = \frac{\theta^{2m}}{(\theta^m + x_{i+1}^m)(\theta^m + x_{i+2}^m)} - x_i, \qquad i = 1, 2, 3, 4,$$

where x_i are positive variables ($x_5 = x_1$, $x_6 = x_2$), has been proposed as a model for sequential disinhibition. Find the steady state and determine the value of m at which a Hopf bifurcation occurs when $\theta = 1/4$.

8. Compute the amplitude and period of the oscillation in the time-delay equation (*example 7*) in the limit as $n \to \infty$.

9. Consider the piecewise linear finite difference equation,

$$x_{t+1} = x_t + 0.4, \qquad 0 \leqslant x_t < 0.6,$$

$$x_{t+1} = x_t - 0.2, \qquad 0.6 \leqslant x_t < 0.7,$$

$$x_{t+1} = x_t - 0.6, \qquad 0.7 \leqslant x_t < 1.0.$$

Determine the dynamics, starting from different initial conditions both algebraically and graphically. Are there any stable cycles?

10. Describe the dynamics in the finite difference equation,

$$x_{t+1} = \frac{1 - x_t}{3x_t + 1} \qquad 0 \leqslant x_t \leqslant 1.$$

Are there any stable cycles?

11. The finite difference equation,

$$x_{t+1} = 3.6x_t - x_t^2,$$

is numerically iterated and gives dynamics that appear to be chaotic, starting from an initial value x_0, $0 < x_0 < 3.6$. After many iterations, what are the maximum and minimum values of x_t that could be observed? (Hint: The values 3.6 and 0 are not the answers.)

12. For the finite difference equation,

$$x_{t+1} = \lambda x_t(1 - x_t), \qquad 0 \leqslant \lambda \leqslant 4, \qquad 0 \leqslant x_0 \leqslant 1,$$

find the values of λ for which a stable, period 2 cycle is found.

13. For the cubic map,

$$x_{t+1} = ax_t^3 + (1 - a)x_t, \qquad -1 \leqslant x_t \leqslant 1, \qquad 0 \leqslant a \leqslant 4,$$

describe the bifurcations and the steady states and cycles for $0 \leqslant a \leqslant 1 + 5^{1/2}$.

14. Consider the simple model for a limit cycle oscillation in equation (2.4). This equation is perturbed by a horizontal translation by an amount b, and there is a rapid relaxation to the limit cycle ($a \to \infty$). (A) Analytically determine the new phase as a function of the old phase (i.e., the PTC) and draw the graphs for $b = 0.8$ and $b = 1.2$. (B) Using the theory in section 7.4 calculate the boundary of the $1:1$ entrainment zone as a function of b. What types of bifurcations occur on the boundary?

Notes and References, Mathematical Appendix

In recent years there has been great interest in nonlinear mathematics and a number of texts in this area have appeared. Physically oriented texts, which contain numerous examples of applications of the theory in the physical sciences but do not contain rigorous mathematics include (Bergé, Pomeau, and Vidal (1984), Schuster (1984), and Thompson and Stewart (1986). Mathematically oriented texts that take a more rigorous approach are Arnold (1983), Guckenheimer and Holmes (1983), and Devaney (1986). An examination of chaotic dynamics from the standpoint of ergodic theory is in Lasota and Mackey (1985). Mathematical treatment of some of the topics included here can be found in recent texts in mathematical biology (Segel 1984; Murray 1988).

A.1 Differential Equations
An excellent introductory text in qualitative theory of differential equations from the eyes of two leading mathematicians is Hirsch and Smale (1974). This contains an elementary discussion of the Hopf bifurcation theorem and a proof of uniqueness and stability limit cycles in the van der Pol equation. Several additional papers on the Hopf bifurcation are in Marsden and McCracken (1976). The Poincaré Index Theorem is usually treated at advanced levels in mathematics. For a discussion of its extension to vector fields in different dimensions and topologies, see Guillemin and Pollack (1974) (look for the Poincaré-Hopf Theorem—but don't assume that this is the same Hopf as in the Hopf bifurcation, for it is not). Applications of the Poincaré-Hopf Index Theorem to biology and chemistry are in Glass (1975).

The phase-plane portrait, which is found for mutual inhibition (figure A.3), captures the topology of the competitive exclusion principle in ecology (May 1973) and serves as a model of mutual inhibition in biochemical and neural networks (Glass and Kauffman 1973; Shymko and Glass 1974; Glass and Young 1979). Dynamics in networks in which there is mutual activation (see above references and also recent studies by Hopfield (1984) display the same topology. Important early examples of chaos in nonlinear ordinary differential equations are in Lorenz (1963) and Rössler (1979).

Applications of time-delay equations as models for feedback control in physiology have been pursued vigorously (Mackey and Glass 1977; an der Heiden 1979, 1985; Mackey 1978; Glass and Mackey 1979a; Mackey 1979a,b; Mackey and an der Heiden 1984; an der Heiden and Mackey, 1982, 1988).

A.2 Finite Difference Equations

A discussion of bifurcations in the quadratic map, (equation A.33) can be found in many places, and the presentations of Devaney (1986) and Thompson and Stewart (1986) are particularly recommended. A good review of the properties of invertible circle maps is in Devaney (1986). In recent years there has been interest in the transition from invertibility to noninvertibility (Feigenbaum, Kadanoff, and Shenker 1982; Ostlund et al. 1983; M. H. Jensen, Bak, and Bohr 1984). From a biological point of view, it is also of interest to examine the dynamics for parameter values for which the map is noninvertible (see section 7.4). Studies of the bifurcations of this map in the noninvertible region were motivated by biological problems (R. Perez and Glass 1982; Glass and Perez 1982) and have subsequently been carried through by many others (Schell, Fraser, and Kapral 1983; Boyland 1986; Fraser and Kapral 1984; Bélair and Glass 1985; and Mackay and Tresser 1986).

References

The most relevant sections for each reference are indicated in brackets.

Abraham, N. B., Gollub, J. P., and Swinney, H. L. 1984. Testing nonlinear dynamics. *Physica* 11D: 252–64. [3.3]

Abrams, P., Feneley, R., and Torrens, M. 1983. *Urodynamics.* Berlin: Springer-Verlag. [1.2]

Adam, D. R., Smith, J. M., Akselrod, S., Nyberg, S., Powell, A. O., and Cohen, R. J. 1984. Fluctuations in T-wave morphology and susceptibility to ventricular fibrillation. *J. Electrocardiol.* 17: 209–18. [8.5]

Adrian, E. D. 1933. Afferent impulses in the vagus and their effects on respiration. *J. Physiol. London* 79: 322–58. [6.1]

Agladze, K. I., and Krinsky, V. I. 1982. Multi-armed vortices in an active chemical medium. *Nature* 296: 424–26. [8.3]

Aihara, K., Numajiri, T., Matsumoto, G., and Kotani, M. 1986. Structures of attractors in periodically forced neural oscillators. *Phys. Lett. A* 116: 313–17. [3.4, 7.4]

Akselrod, S., Gordon, D., Ubel, A., Shannon, D.C., Barger, A.C., and Cohen, R. J. 1981. Power spectrum analysis of heart-rate fluctuation: A quantitative probe of beat-to-beat cardiovascular control. *Science* 213: 220–22. [3.3, 9.1]

Allen, T. T. 1983. On the arithmetic of phase locking: Coupled oscillators as a lattice on R². *Physica* 6D: 305–21. [7.3]

Allesssie, M. A., Bonke, F.I.M., and Schopman, F.J.G. 1977. Circus movement in rabbit atrial muscle as a mechanism of tachycardia. III. The "leading circle" concept: A new model of circus movement in cardiac tissue without the involvement of an anatomic obstacle. *Circ. Res.* 41: 9–18. [8.3, 8.5]

Allessie, M. A., Lammers, W.J.E.P., Bonke, F.I.M., and Hollen, J. 1985. Experimental evaluation of Moe's multiple wavelet hypothesis of atrial fibrillation. In *Cardiac Electrophysiology and Arrythmias,* ed. D. P. Zipes and J. Jalife. Orlando: Grune and Stratton. [8.5]

Arnold, V. I. 1965, Small denominators. I. Mappings of the circumference onto itself. *Am. Math. Soc. Translations,* ser. 2, vol. 46, pp. 213–84. Providence: American Mathematics Soceity. [7.2]

———. 1983. *Geometrical Methods in the Theory of Ordinary Differential Equations.* New York: Springer-Verlag. [2.4, 5.2, 5.3, 7.2, A.1, A.2]

Arvanitaki, A. 1939. Recherches sur la réponse oscillatoire locale de l'axone géant isolé de "Sepia". *Arch. Int. Physiol.* 49: 209–56. [5.1]

Atwater, I., Dawson, C. M., Scott, A., Eddlestone, G., Rojas, E. 1980. The nature of oscillatory behavior in electrical activity for pancreatic β-cell. In *Biochemistry and Biophysics of the Pancreatic-β-Cell*, pp. 100–107. Stuttgart: Georg Thieme Verlag. [1.1]

Ayers, A. L., and Selverston, A. I. 1979. Monosynaptic entrainment of an endogenous pacemaker network: A cellular mechanism for von Holst's magnet effect. *J. Comp. Physiol.* 129: 5–17. [7.1]

Babloyantz, A., and Destexhe, A. 1986. Low-dimensional chaos in an instance of epilepsy. *Proc. Natl. Acad. Sci.* 83: 3513–17. [3.4]

Baconnier, P., Benchetrit, G., Demongeot, J., and Pham Dinh, T. 1983. Simulation of the entrainment of the respiratory rhythm by two conceptually different models. In *Rhythms in Biology and Other Fields of Application, Lecture Notes in Biomathematics, vol. 49*, ed. M. Cosnard, J. Demongeot, and A. Le Breton, pp. 2–16. Berlin: Springer-Verlag. [6.2, 7.3]

Battersby, E. J. 1965. Pacemaker periodicity in atrial fibrillation. *Circ. Res.* 17: 296–302. [8.5]

Bélair, J. 1986. Periodic pulsatile stimulation of a nonlinear oscillator. *J. Math. Biol.* 24: 217–32. [7.3]

Bélair, J., and Glass, L. 1985. Universality and self-similarity in the bifurcations of circle maps. *Physica* 16D: 143–54. [A.2]

Bellett, S. 1971. *Clinical Disorders of the Heartbeat.* Philadelphia: Lea & Febiger. [7.5, 8.1, 9.1]

Benchetrit, G., Baconnier, P., and Demongeot, J., eds. 1987. *Concepts and Formalizations in the Control of Breathing.* Manchester: Manchester University Press. [1.1]

Bennett, M. A., and Pentecost, B. L. 1970. The pattern of onset and spontaneous cessation of atrial fibrillation in man. *Circulation* 41: 981–88. [8.5]

Bergé, P., Pomeau, Y., and Vidal, C. 1984. *Order Within Chaos: Towards a Deterministic Approach to Turbulence.* Paris: Johny Wiley and Hermann. [A.1, A.2]

Bernard, C. 1878. *Leçons sur les Phenomènes de la Vie Commun aux Animaux et aux Vegetaux.* Paris: Balliere. [1.1]

Berridge, M. J., and Rapp, P. E. 1979. A comparative survey of the function, mechanism and control of cellular oscillators. *J. Exp. Biol.* 81: 217–79. [4.1]

Best, E. N. 1979. Null space in the Hodgkin-Huxley equations: A critical test. *Biophys. J.* 27: 87–104. [5.3, 6.5]

Bickers, W. 1941. Uterine contractions in dysmenorrhea. *Am. J. Obstet. Gynecol.* 42: 1023–30. [5.2]

Bortoff, A. 1961. Electrical activity of intestine recorded with pressure electrode. *Am. J. Physiol.* 201: 209–12. [5.1]

Boyett, M. R., and Jewell, B. R. 1978. A study of the factors responsible for the rate-dependent shortening of the action potential in mammalian ventricular muscle. *J. Physiol.* 285: 359–80. [8.1]

Boyland, P. 1986. Bifurcations of circle maps: Arnold tongues, bistability and rotation intervals. *Commun. Math Phys.* 106: 353–381.

Bradley, G. W., Euler, C. von, Martilla, I., and Roos, B. 1975. A model of the central and reflex inhibition of inspiration in the cat. *Biol. Cybern.* 19: 105–16. [6.2]

Bramble, D. M. 1983. Respiratory patterns and control during unrestrained human running. In *Modelling and Control of Breathing*, ed. B. J. Whipp and D. M. Wiberg, pp. 213–20. New York: Elsevier. [7.5]

Bramble, D. M., and Carrier, D. R. 1983. Running and breathing in mammals. *Science* 219: 251–56. [7.5]

Breuer, J. 1868. Die Selbststeuerung der Athmung durch den Nervus Vagus. *Sitzber. Akad. Wiss. Wien* 58: 909–37 (English trans.) by E. Ullman, 1970). In *Breathing: Hering-Breuer Centenary Symposium*, ed. R. Porter, pp. 365–94. London: Churchill. [6.1]

Brink, F., Bronk, D. W., and Larrabee, M. G. 1946. Chemical excitation of nerve. *Ann. N.Y. Acad. Sci.* 47: 457–85. [5.1]

Brown, B. H., Duthie, H. L., Horn, A. R., and Smallwood, R. H. 1975. A linked oscillator model of electrical activity of human small intestine. *Am. J. Physiol.* 229: 384–88. [8.1]

Brown, T. G. 1914. On the nature of the fundamental activity of the nervous centres; together with an analysis of the conditioning of rhythmic activity in progression, and a theory of the evolution of function in the nervous system. *J. Physiol. (Lond.)* 48: 18–46. [4.2, 4.3]

Builder, G., and Roberts, N. F. 1939. The synchronisation of a simple relaxation oscillation. *Australasia Wireless Association Tech. Rev.* 4: 165–80. [7.3]

Burch, P.R.J. 1976. *The Biology of Cancer: A New Approach.* Baltimore: University Park Press. [3.2]

Burns, B. D., and Salmoiraghi, G. C. 1960. Repetitive firing of respiratory neurons, during their burst activity. *J. Neurophysiol.* 23: 27–46. [4.3]

Butler, P. J., and Woakes, A. J. 1980. Heart-rate, respiratory frequency and wing beat frequency of free flying barnacle geese, branta-leucopsis. *J. Exp. Biol.* 85: 213–26. [7.5]

Cannon, W. B. 1926. Physiological regulation of normal states: Some tentative postulates concerning biological homeostatics. In *A*

Charles Richet: Ses Amis, Ses Colleagues, Ses Elèves, ed. A. Pettit, pp. 91–93. Paris: Les Edition Medicale. [1.1]

———. 1929. Organization for physiological homeostasis. *Physiol. Rev.* 9: 399–431. [1.1]

Carpenter, G. A. 1979. Bursting phenomena in excitable membranes. *SIAM J. Appl. Math.* 36: 334–72. [8.1]

Cartwright, M. L., and Littlewood, J. E. 1945. On non-linear differential equations of the second order. I. The equation $\ddot{y} - k(1 - y^2)/\dot{y} + y = b\lambda k \cos(\lambda t + \alpha)$, k large. *J. London Math Soc.* 20: 180–89. [7.2]

Castellanos, A., Luceri, R. M., Moleiro, F., Kayden, D. S., Trohman, R. G., Zaman, L., and Myerburg, R. J. 1984. Annihilation, entrainment and modulation of ventricular parasystolic rhythms. *Am. J. Cardiol.* 54: 317–22. [5.4, 6.4, 7.5]

Chalazontis, N., and Boisson, M., eds. 1978. *Abnormal Neuronal Discharges*. New York: Raven Press. [4.1]

Chay, T. R. 1984. Abnormal discharges and chaos in neuronal model system. *Biol. Cybern.* 50: 301–11. [4.1]

Chay, T. R., and Lee, Y. S. 1984. Impulse responses of automaticity in the Purkinje fiber. *Biophys. J.* 45: 841–49. [6.5]

———. 1985. Phase resetting and bifurcation in the ventricular myocardium. *Biophys. J.* 47: 641–51. [6.5]

Chay, T. R., and Rinzel, J. 1985. Bursting, beating and chaos in an excitable membrane model. *Biophys. J.* 45: 357–66. [2.1, 4.1]

Cherniack, N. S., Euler, C. von, Homma, I., and Kao, F. F. 1979. Graded changes on the ventral surface of the medulla. *J. Physiol. (Lond.)* 287: 191–211. [5.2]

Christensen, J., and Wingate, D. L. 1983. *A Guide to Gastrointestinal Motility*. Bristol: Wright. [1.1]

Clark, F. J., and Euler, C. von. 1972. On the regulation of depth and rate of breathing. *J. Physiol. (Lond.)* 222: 267–95. [6.1]

Clay, J. R., Guevara, M. R., and Shrier, A. 1984. Phase resetting of the rhythmic activity of embryonic heart cell aggregates: Experiment and theory. *Biophys. J.* 45: 699–714. [2.3, 6.5]

Cohen, A. H., and Wallen, P. 1980. The neuronal correlate of locomotion in fish. *Exp. Brain Res.* 41: 11–8. [5.2]

Cohen, A. H., Rossignol, S., and Grillner, S., eds. 1988. *Neural Control of Rhythmic Movements in Vertebrates*. New York: John Wiley. [1.1, 4.2]

Cohen, D. S., Neu, J. C., and Rosales, R. R. 1978. Rotating spiral wave solutions of reaction-diffusion equations. *SIAM J. Appl. Math.* 35: 536–47. [8.3]

Cohen, M. I. 1974. The genesis of respiratory rhythmicity. In *Central-*

Rhythmic and Regulation, ed. W. Umbach and H. P. Koepchen, pp. 15–35. Stuttgart: Hippokrates. [4.4]

———. 1979. Neurogenesis of respiratory rhythm in the mammal. *Physiol. Rev.* 59: 1105–73. [4.2]

Cohen, M. I., and Feldman, J. L. 1977. Models of respiratory phase switching. *Federation Proc.* 36: 2367–74. [6.2]

Constantinou, C. E. 1974. Renal pelvic pacemaker control of ureteral peristaltic rate. *Amer. J. Physiol.* 226: 1413–19. [1.1]

Cooley, J., Dodge, F., and Cohen, H. 1965. Digital computer solutions for excitable membrane models. *J. Cell Comp. Physiol.* 66: 99–108. [5.3]

Crammer, L. 1959. Periodic psychoses. *Br. Med. J.* 1: 545–49. [9.1]

———. 1960. All cycles are not menstrual. *Lancet* 2: 874–75. [9.1]

Cronin-Scanlon, J. 1974. A mathematical model for catatonic schizophrenia. *Ann. N.Y. Acad. Sci.* 231: 112–22. [4.5]

Crutchfield, J., Farmer, D., Packard, N., Shaw, R., Jones, G., and Donnelly, R. J. 1980. Power spectral analysis of a noisy system. *Phys. Lett.* 76A: 1–4. [3.3]

Curzi-Dascalova, L., Radvanyi, M. F., Moriette, G., Morel-Kahn, F., and Korm, G. 1979. Respiratory variability according to sleep states during mechanical ventilation: A polygraphic study in a baby with bilateral diaphragmatic paralysis. *Neuropaediatrie* 10: 361–69. [7.5]

Cvitanovic, P., ed. 1984. *Universality in Chaos.* Bristol: Adam Hilger. [3.3]

Czeisler, C. A., Allan, J. S., Strogatz, S. H., Ronda, J. M., Sanchez, R., Rios, C. D., Freitag, W. O., Richardson, G. S., and Kronauer, R. E. 1986. Bright light resets the human circadian pacemaker independent of the timing of the sleep-wake cycle. *Science* 233: 667–71. [7.5, 9.4]

Daan, S., and Beersma, D. 1984. Circadian gating of human sleep and wakefulness. In *Mathematical Modelling of Circadian Systems*, ed. M. Moore-Ede and C. A. Czeisler. New York: Raven Press. [7.3, 7.5]

Daan, S., Beersma, D.G.M., and Borbely, A. A. 1984. Timing of human sleep: Recovery process gated by a circadian pacemaker. *Am. J. Physiol.* 246 (*Regulatory Integrative Comp. Physiol.* 15): R161–78. [7.3, 7.5]

Dale, D. C., Alling, D. W., and Wolff, S. M. 1972. Cyclic hematopoiesis: The mechanism of cyclic neutropenia in grey collie dogs. *J. Clin. Invest.* 51: 2197–2204. [9.3]

Daly, D. C., Soto-Albors, C., Walters, C., Ying, Y.-K., and Riddick, D. H. 1985. Ultrasonographic assessment of luteinized unruptured follice syndrome in unexplained infertility. *Fertil Steril.* 43: 62–65. [5.1]

Decherd, G. M., and Ruskin, A. 1946. The mechanism of the Wencke-bach type of A-V block. *Br. Heart J.* 8: 6–16. [8.1]

DeHaan, R. L. 1967. Regulation of spontaneous activity and growth of embryonic chick heart cells in tissue culture. *Devel. Biol.* 23: 226–40. [2.3]

DeHaan, R. L., and Fozzard, H. A. 1975. Membrane response to current pulses in spheroidal aggregates of embryonic heart cells. *J. Gen. Physiol.* 65: 207–22. [2.3]

Dellow, P. G., and Lund, J. P. 1971. Evidence for central timing of rhythmical mastication. *J. Physiol. (Lond.)* 215: 1–13. [5.3]

Denjoy, A. 1932. Sur les courbes définies par les équations differentielle à la surface du tore. *J. Math.* 11, fasc. IV: 333–75. [7.2]

Devaney, R. L. 1986. *An Introduction to Chaotic Systems.* Menlo Park: Benjamin/Cummings. [7.2, A.1, A.2]

Diamant, N. E.; Rose, P. K.; and Davison, E. J. 1970. Computer simulation of intestinal slow-wave frequency gradient. *Am. J. Physiol.* 219: 1684–90. [8.1]

DiFrancesco, D. 1984. Characterization of the pace-maker current kinetics in calf Purkinje fibres. *J. Physiol. (Lond.)* 348: 341–67. [4.1]

DiFrancesco, D., and Noble, D. 1985. A model of cardiac electrical activity incorporating ionic pumps and concentration changes, *Phil. Trans. R. Soc. Lond.* B. 307: 353–98. [4.1]

Dowell, A. R., Buckley, E., Cohen, R., Whalen, R. E., and Sieker, H. O. 1971. Cheyne-Stokes respiration: A review of clinical manifestations and critique of physiological mechanism. *Arch. Int. Med.* 127: 712–26. [4.5]

Downar, E., Parson, I. D., Mickleborough, L. L., Cameron, D. A., Yao, L. C., and Waxman, M. B. 1984. On-line epicardial mapping of intraoperative ventricular arrhythmias: Initial clinical experience. *J. Amer. Coll. Cardiol.* 4: 703–14. [8.5]

Durston, A. J. 1973. Dictyostelium discoideum aggregation fields as excitable media. *J. Theor. Biol.* 42: 483–504. [8.3]

Dvorak, I., and Siska, J. 1986. On some problems encountered in the estimation of the correlation dimension of the EEG. *Phys. Lett. A* 118: 63–66. [3.4]

Eckmann, J.-P. 1981. Roads to turbulence in dissipative dynamical systems. *Rev. Mod. Phys.* 53: 643–54. [3.3]

Eckmann, J.-P., and Ruelle, D. 1985. Ergodic theory of chaos and strange attractors. *Rev. Mod. Phys.* 57: 617–56. [3.4]

El-Sherif, N. 1985. The figure 8 model of reentrant excitation in the canine post infarction heart. In *Cardiac Electrophysiology and Arrhythmias*, ed. D. P. Zipes and J. Jalife, pp. 363–78. Orlando: Grune and Stratton. [8.5]

Erneux, T., and Herschkowitz-Kaufman, M. 1975. Rotating waves as asymptotic solutions of a model chemical reaction. *J. Chem. Phys.* 66: 248–50. [8.2]

Euler, C. von. 1986. Brain-stem mechanisms for generation and control of the breathing pattern. In *Handbook of Physiology: The Respiratory System*, sec. 3, vol. 2, ed. N. S. Cherniack and J. G. Widdicombe, pp. 1–67. Bethesda: American Physiological Society. [4.4]

Fallert, M., and Mühlemann, R. 1971. Der Hering-Breuer Reflex bei künstlicher Beatmung des Kaninchens. I. Die Auslösung der reflektorischen Inspirationen durch den Respirator. *Pfluegers Arch.* 330: 162–74. [7.1]

Farley, B. G., and Clark, W. A. 1961. Activity in networks of neuron-like elements. In *Information Theory*, 4th London Symposium, ed. C. Cherry. London: Butterworths. [8.3]

Farmer, J. D. 1982. Chaotic attractors of an infinite-dimensional dynamical system *Physica* 4D: 366–93. [4.6]

Farmer, J. D., Ott, E., and Yorke, J. A. 1983. The dimension of chaotic attractors. *Physica* 7D: 153–80. [3.4]

Fatt, P., and Katz, B. 1952. Spontaneous subthreshold activity at motor nerve endings. *J. Physiol. (Lond.).* 117: 109–28. [3.1]

Feigenbaum, M. J. 1978. Quantitative universality for a class of nonlinear transformations. *J. Stat. Phys.* 19: 25–52. [2.5]

Feigenbaum, M., Kadanoff, L. P., and Shenker, S. 1982. Quasiperiodicity in dissipative systems: A renormalization group analysis. *Physica* 5D: 370–86. [A.2]

Feldman, J. L. 1986. Neurophysiology of breathing in mammals. In *Handbook of Physiology—The Nervous System. IV: Intrinsic Regulatory Systems of the Brain*, ed. F. E. Bloom, pp. 463–524. Bethesda: American Physiological Society. [4.4, 6.1]

Feldman, J. L., and Cowan, J. D. 1975. Large-scale activity in neural nets. II: A model for the brainstem respiratory oscillator. *Biol. Cybern.* 17: 39–51. [6.2]

Feller, W. 1968. *An Introduction to Probability Theory and Its Applications*, vols. I, II, 3d ed. New York: John Wiley. [3.1]

Feroe, J. A. 1983. Traveling waves with finitely many pulses in a nerve equation. In *Lectures Notes in Biomathematics 51: Oscillations in Mathematical Biology*, ed. J.P.E. Hodgson, pp. 61–101. Berlin: Springer-Verlag. [8.1]

Fincham, W., and Liassides, C. 1978. The role of the van der Pol oscillator in the control of breathing in the human. In *Proceedings of the UK Simulation Council Conference on Computer Simulation.* Surrey, U.K.: IPC Science and Technology Press. [6.2]

Findley, L. J., and Capildeo, R., eds. 1984. *Movement Disorders: Tremor.* London: Macmillan. [1.1, 3.3]

Flaherty, J. E., and Hoppensteadt, F. C. 1978. Frequency entrainment of a forced van der Pol oscillator. *Studies in Appl. Math.* 58: 5–15. [7.2]

Fohlmeister, J. F., Poppele, R. E., and Purple, R. L. 1974. Repetitive firing: Dynamic behavior of sensory neurons reconciled with a quantitative model. *J. Neurophysiol.* 37: 1213–27. [7.3]

Forssberg, H. S., Griller, S., Rossignol, S., and Wallen, P. 1976. Phase control of reflexes during locomotion in vertebrates. In *Neural Control of Locomotion*, ed. R. M. Herman, S. Grillner, P.S.G. Stein, and D. G. Stuart, pp. 647–74. New York: Plenum. [6.4]

Fraser, S., and Kapral, R. 1984. Universal vector scaling in one-dimensional maps. *Phys. Rev. A* 30: 1017–25. [A.2]

Furuse, A., Matsuo, H., and Saigusa, M. 1981. Effects of intervening beats on ectopic cycle lengthy in a patient with ventricular parasystole. *Jpn. Heart J.* 22: 201–209. [6.4]

Gallagher, J. J. 1985. Variants of preexcitation: Update 1984. In *Cardiac Electrophysiology and Arrhythmias*, ed. D. P. Zipes and J. Jalife, pp. 419–33. Orlando: Grune and Stratton. [8.2]

Gander, P. H., Kronauer, R. E., Czeisler, C. A., and Moore-Ede, M. C. 1984. Simulating the action of zeitgebers on a coupled two-oscillator model of the human circadian system. *Am. J. Physiol.* 247 (*Regulatory Integrative Comp. Physiol.* 16): R418–26. [7.5]

Garrey, W. E. 1914. Nature of fibrillary contraction in the heart. *Amer. J. Physiol.* 33: 397–414. [8.2, 8.3, 8.5]

———. 1924. Auricular fibrillation. *Physiol. Rev.* 4: 215–50. [8.2, 8.3, 8.5]

Gatti, R. A., Robinson, W. A., Deinare, A. S., Nesbit, M., McCullogh, J. J., Ballow, M., and Good, R. A. 1973. Cyclic leukocytosis in chronic myelogenous leukemia. *Blood* 41: 771–82. [1.2]

Gavosto, F. 1974. Granulopoiesis and cell kinetics in chronic myeloid leukemia. *Cell Tissue Kinet.* 7: 151–63. [4.6]

Geman, S., and Miller, M. 1976. Computer simulation of brain stem respiratory activity. *J. Appl. Physiol.* 41: 931–38. [6.2]

Gerisch, G. 1965. Stadienspezifische Aggregationmuster bei *Dictyostelium discoideum. Wilhelm Roux Archiv Entwicklungsmech. Organismen* 156: 127–44. [8.3]

Gerstein, G. L., and Mandelbrot, M. 1964. Random walk models for the spike activity of a single neuron. *Biophys. J.* 4: 41–68. [3.1]

Glass, L. 1975. A topological theorem for nonlinear dynamics in chemical and ecological networks. *Proc. Nat. Acad. Sci. USA* 72: 2856–57. [A.1]

————. 1987. Is the respiratory rhythm generated by a limit cycle oscillation? In *Concepts and Formalizations in the Control of Breathing*, ed. G. Benchetrit, P. Baconnier, and J. Demongeot, pp. 247–63. Manchester: Manchester University Press. [5.2, 5.3]

Glass, L., and Bélair, J. 1986. Continuation of Arnold tongues in mathematical models of periodically forced biological oscillators. In *Nonlinear Oscillations in Biology and Chemistry*, ed. H. G. Othmer, pp. 232–43. Berlin: Springer-Verlag. [7.3, 7.4]

Glass, L., and Kauffman, S. A. 1973. The logical analysis of continuous nonlinear biochemical control networks. *J. Theor. Biol.* 39: 103–29.

Glass, L., and Mackey, M. C. 1979a. Pathological conditions resulting from instabilities in physiological control systems. *Ann. N.Y. Acad. Sci.* 316: 214–35. [4.5, 4.6, A.1]

————. 1979b. A simple model for phase locking of biological oscillators. *J. Math. Biol.* 7: 339–52. [7.3]

Glass, L., and Pasternack, J. S. 1978a. Stable oscillations in mathematical models of biological control systems. *J. Math. Biol.* 6: 207–23. [4.4]

————. 1978b. Prediction of limit cycles in mathematical models of biological oscillations. *Bull. Math. Biol.* 40: 27–44. [4.4, 5.2]

Glass, L., and Perez, R. 1982. The fine structure of phase locking. *Phys. Rev. Lett.* 48: 1772–75. [A.2]

Glass, L., and Winfree, A. T. 1984. Discontinuities in phase-resetting experiments. *Am. J. Physiol.* 246 (*Regulatory Integrative Comp. Physiol.* 15): R251–58. [5.4, 6.3]

Glass, L., and Young, R.E. 1979. Structure and dynamics of neural network oscillators. *Brain Res.* 179: 207–18. [4.4, A.1]

Glass, L., Beuter, A., and Larocque, D. 1988. Time delays, oscillations and chaos in physiological control systems. *Math. Biosci.* In press. [4.5, 9.1]

Glass, L., Guevara, M. R., and Shrier, A. 1987. Universal bifurcations and the classification of cardiac arrhythmias. *Ann. N.Y. Acad. Sci.* 504: 168–178. [7.1, 8.1]

Glass, L., Shrier, A., and Bélair, J. 1986. Chaotic cardiac rhythms. In *Chaos*, ed. A. Holden, pp. 237–56. Manchester: Manchester University Press. [1.3]

Glass, L., Graves, C., Petrillo, G. A., and Mackey, M. C. 1980. Unstable dynamics of a periodically driven oscillator in the presence of noise. *J. Theor. Biol.* 86: 455–75. [7.3]

Glass, L., Guevara, M. R., Shrier, A., and Perez, R. 1983. Bifurcation and chaos in a periodically stimulated cardiac oscillator. *Physica* 7D: 89–101. [7.1, 7.4, A.2]

Glass, L., Guevara, M. R., Bélair, J., and Shrier, A. 1984. Global bifurcations of a periodically forced biological oscillator. *Phys. Rev.* 29: 1348–57. [1.3, 7.1, 7.4]

Glass, L., Goldberger, A. L., Courtemanche, M., and Shrier, A. 1987. Nonlinear dynamics, chaos and cardiac arrhythmias. *Proc. R. Soc. Lond.* A 413: 9–26.

Gleick, J. 1987. *Chaos: Making a New Science.* New York: Viking [1.1]

Goldberger, A. L., and Goldberger, E. 1986. *Clinical Electrocardiography: A Simplified Approach,* ed. 3 St. Louis: C. V. Mosby. [1.1, 9.1]

Goldberger, A. L., and Rigney, D. R. 1988. Sudden death is not chaos. In *Dynamic Patterns in Complex Systems,* ed. J. A. S. Kelso, M. F. Schlesinger, and A. J. Mandell. Singapore: World Scientific. [9.1]

Goldberger, A. L., Bhargava, V., West, B. J., and Mandell, A. J. 1985. On a mechanism of cardiac electrical stability: The fractal hypothesis. *Biophys. J.* 48: 525–28. [3.4, 9.1]

———. 1986. Some observations on the question: Is ventricular fibrillation "chaos"? *Physica* 19D: 282–89. [8.5, 9.1]

Goodman, L. 1964. Oscillatory behavior of ventilation in resting man. *IEEE Trans. Biomed. Eng.* BME-11: 82–93. [3.3, 9.1]

Grassberger, P., and Procaccia, I. 1983. Measuring the strangeness of strange attractors. *Physica* 9D: 189–208. [3.4, 4.6]

Graves, C., Glass, L., Laporta, D., Meloche, R., and Grassino, A. 1986. Respiratory phase locking during mechanical ventilation in anesthetized human subjects. *Am. J. Physiol.* 250 (*Regulatory, Integrative Comp. Physiol.* 19): R902–909. [7.5]

Grebogi, C., Ott, E., Pelikan, S., and Yorke, J. A. 1984. Strange attractors that are not chaotic. *Physica* 13D: 261–68. [3.4]

Grebogi, C., McDonald, S. W., Ott, E., and Yorke, J. A. 1985. Exterior dimension of fat fractals. *Phys. Lett.* 110A: 1–4. [3.4]

Greenberg, J. M., Hassard, B. D., and Hastings, S. P. 1978. Pattern formation and periodic structures in systems modeled by reaction-diffusion equations. *Bull. Am. Math. Soc.* 84: 1296–327. [8.3]

Greenberg, M. L., Channano, A. D., Cronkite, E. P., Giacomelli, G., Rai, K. R., Schiffer, L. M., Stryckmons, P. A., and Vincent, P. C. 1972. The generation time of human leukemic myeloblasts. *Lab. Invest* 26: 245–52. [4.6]

Grillner, S. 1981. Control of locomotion in bipeds, tetrapods and fish. In *Handbook of Physiology, Motor Control,* ed. V. Brooks, pp. 1179–236. [5.2]

Grillner, S., and Wallen, P. 1984. How does the lamprey central nervous system make the lamprey swim? *J. Exp. Biol.* 112: 337–57. [5.2]

Grodins, F. S. 1963. *Control Theory and Biological Systems.* New York: Columbia University Press. [4.5, 9.2]

Grossmann, S., and Thomae, S. 1977. Invariant distributions and stationary correlation functions of one-dimensional discrete processes. *Z. Naturforsch.* 32a: 1353–63. [2.5]

Guckenheimer, J. 1975. Isochrons and phaseless sets. *J. Math. Biol.* 1: 259–73. [6.3]

———. 1982. Noise in chaotic systems. *Nature* 298: 358–61. [3.3]

Guckenheimer, J., and Holmes, P. 1983. *Nonlinear Oscillations, Dynamical Systems and Bifurcations of Vector Fields.* New York: Springer-Verlag. [3.3, 5.2, 5.3, 7.2, A.1, A.2]

Guevara, M. R. 1984. Chaotic Cardiac Dynamics. Ph.D. dissertation, McGill University. [7.1, 7.4, 9.4]

———. 1987. Afterpotentials and pacemaker oscillations in an ionic model of cardiac Purkinje fibre. In *Temporal Disorder in Human Oscillatory Systems,* ed. L. Rensing, U. an der Heiden, and M. C. Mackey, pp. 126–33. Berlin: Springer-Verlag. [5.1]

Guevara, M. R., and Glass, L. 1982. Phase locking, period doubling bifurcations and chaos in a mathematical model of a periodically driven oscillator: A theory for the entrainment of biological oscillators and the generation of cardiac dysrhythmias. *J. Math. Biol.* 14: 1–23. [2.3, 7.4, A.3]

Guevara, M. R., Glass, L., and Shrier, A. 1981. Phase locking, period-doubling bifurcations, and irregular dynamics in periodically stimulated cardiac cells, *Science* 214: 1350–53. [1.3, 2.3, 7.1, 7.4]

Guevara, M. R., Shrier, A., and Glass, L. 1986. Phase resetting of spontaneously beating embryonic ventricular heart-cell aggregates. *Amer. J. Physiol.* 251 (*Heart Circ. Physiol.* 20): H1298–1305. [5.1, 6.5]

Guevara, M. R., Shrier, A., and Glass, L. 1988. Phase-locked rhythms in periodically stimulated heart cell aggregates. *Amer. J. Physiol.* 254 (*Heart Circ. Physiol.* 23): H1–10. [7.1, 9.4]

Guevara, M. R., Glass, L., Mackey, M. C., and Shrier, A. 1983. Chaos in neurobiology. *IEEE Trans. Syst. Man Cybern.* SMC-13: 790–98. [7.4, 9.1]

Guevara, M. R., Ward, G., Shrier, A., and Glass, L. 1984. Electrical alternans and period-doubling bifurcations. In *Computers in Cardiology,* pp. 167–70. Long Beach, Calif.: IEEE Computer Society. [8.1]

Guillemin, V., and Pollack, A. 1974. *Differential Topology*. Englewood Cliffs, N.J.: Prentice-Hall. [A.1]

Gulko, F. B., and Petrov, A. A. 1972. Mechanism of formation of closed pathways of conduction in excitable media. *Biofizika* 15: 513–20. [8.3]

Gurney, C. W., Simmons, E. L., and Gaston, E. O. 1981. Cyclic erythropoiesis in W/W^v mice following a single small dose of ^{89}Sr. *Exp. Hematol.* 9: 118–22. [9.4]

Guttman, R., and Barnhill, R. 1970. Oscillation and repetitive firing in squid axons. *J. Gen. Physiol.* 55: 104–18. [5.1]

Guttman, R., Feldman, L., Jakobsson, E. 1980. Frequency entrainment of squid axon membrane. *J. Memb. Biol.* 56: 9–18. [7.1]

Guttman, R., Lewis, S., and Rinzel, J. 1980. Control of repetitive firing in squid axon membrane as a model for a neuron oscillator. *J. Physiol. (Lond.)* 305: 377–95. [5.1, 5.3, 5.4]

Guyton, A. C., Crowell, J. W., and Moore, J. W. 1956. Basic oscillating mechanism of Cheyne-Stokes breathing. *Am. J. Physiol.* 187: 395–98. [9.3]

Hagan, P. S. 1982. Spiral waves in reaction-diffusion equations. *SIAM J. Appl. Math.* 42: 762–86. [8.3]

Hao, B.-L., ed. 1984. *Chaos*. Singapore: World Scientific. [3.3]

Harker, G.F.H. 1938. The mechanism of synchronization in the linear time base. *Phil. Mag. 7 Series* 26: 193–213. [7.3]

Harmon, L. D. 1964. Neuromines: Action of a reciprocally inhibitory pair. *Science* 146: 1323–25. [4.3]

Hayashi, C. 1964. *Nonlinear Oscillations in Physical Systems*. New York: McGraw-Hill. Reprinted by Princeton University Press (1985). [7.2]

Hayes, N. D. 1950. Roots of the transcendental equation associated with a certain difference-differential equation. *J. Lond. Math. Soc.* 25: 226–32. [A.1]

an der Heiden, U. 1979. Delays in physiological systems. *J. Math. Biol.* 8: 345–64. [4.6, A.1]

———. 1985. Stochastic properties of simple differential-delay equations In *Delay Equations, Approximation and Application*, G. Meinardus and G. Nürnberger. Basel: Birkhauser. [4.6, A.1]

an der Heiden, U., and Mackey, M. C. 1982. The dynamics of production and destruction: Analytic insight into complex behaviour. *J. Math. Biol.* 16: 75–101. [4.6, A.1]

———. 1988. Dynamics, health, and disease. *SIAM Review*. In Press. [4.6, 9.1, A.1]

an der Heiden, U., Mackey, M. C., and Walther, H. O. 1981. Complex oscillations in a simple deterministic neuronal network. In *Lectures in Applied Mathematics*, ed. F. Hoppensteadt, pp. 355–60. Providence: American Mathematical Society. [4.6]

Herbschleb, J. N., Heethar, R. M., Van der Tweel, I., Zimmerman, A.N.E., and Meijler, F. L. 1979. Signal analysis of ventricular fibrillation. In *Computers in Cardiology*, pp. 49–54. Long Beach, Calif: IEEE Computer Society. [8.5]

Herczynski, R., and Karczewski, W. 1976. Neural control of breathing: A system analysis. *Acta Physiol. Pol.* 27: 109–30. [6.2]

Hirsch, M. W., and Smale, S. 1974. *Differential Equations, Dynamical Systems and Linear Algebra*. New York: Academic. [2.4, A.1]

Hodgkin, A. L., and Huxley, A. F. 1952. A quantitative description of membrane current and its application to conduction and excitation in nerve. *J. Physiol. (Lond.)* 117: 500–44. [2.1, 4.1]

Hohnloser, H., Weiss, M., Zeiher, A., Wollschlarger, H., Hust, M. H., and Just, H. 1984. Sudden cardiac death recorded during ambulatory electrocardiographic monitoring. *Clin. Cardiol.* 7: 517–23. [8.5, 9.1]

Holden, A. V., ed. 1986. *Chaos*. Manchester: Manchester University Press. [1.1]

Holden, A. V., Winlow, W., and Haydon, P. G. 1982. The induction of periodic and chaotic activity in a molluscan neurone. *Biol. Cybern.* 43: 169–73. [4.1, 5.1]

Holst, E. von. 1973. *The Behavioural Physiology of Animals and Man: The Collected Papers of Erich von Holst*, vol. 1. Coral Gables: University of Miami Press. [7.5]

Honerkamp, J. 1983. The heart as a system of coupled nonlinear oscillators. *J. Math. Biol.* 18: 69–88. [7.4, 7.5]

Hopfield, J. J. 1984. Neurons with graded response have collective computational properties like those of two-state neurons. *Proc. Natl. Acad. Sci. USA* 81: 3088–92. [4.3, A.1]

Hoppensteadt, F. C., and Keener, J. 1982. Phase locking of biological clocks. *J. Math. Biol.* 15: 339–49. [2.3, 7.4]

Hosomi, H., and Hayashida, Y. 1984. Systems analysis of blood pressure oscillation. In *Mechanisms of Blood Pressure Waves*, ed. K. Miyakawa, H. P. Koepchen, and C. Polosa, pp. 215–27. Tokyo: Japan Scientific Societies Press. [1.1, 4.5]

Huxley, A. F. 1959. Ion movements during nerve activity. *Ann. N.Y. Acad. Sci.* 81: 221–46. [4.1, 5.1]

Ideker, R. E., and Shibata, N. 1986. Experimental evaluation of predictions about ventricular fibrillation based upon phase resetting (Abstract). Vancouver: Int. Union Physiol. Sciences. [8.5]

Ikeda, N., Tsuruta, H., and Sato, T. 1981. Difference equation model of the entrainment of myocardial pacemakers based on the phase response curve. *Biol. Cybern.* 42: 117–28. [7.4, 7.5]

Iscoe, S., and Vanner, S. 1980. Respiratory periodicity following stimulation of vagal afferents. *Can J. Physiol. Pharmacol.* 58: 823–9. [6.1]

Jalife, J., and Antzelevitch, C. 1979. Phase resetting and annihilation of pacemaker activity in cardiac tissue. *Science* 206: 695–7. [1.3, 6.4, 7.1]

————. 1980. Pacemaker annihilation: Diagnostic and therapeutic implications. *Am. Heart J.* 100: 128–30. [5.1, 5.4]

Jalife, J., and Michaels, D. C. 1985. Phase-dependent interactions of cardiac pacemakers as mechanisms of control and synchronization in the heart. In *Cardiac Electrophysiology and Arrhythmias*, ed. D. P. Zipes and J. Jalife, pp. 109–19. Orlando: Grune and Stratton. [6.1, 6.4, 7.1]

Jalife, J., and Moe, G. K. 1976. Effect of electrotonic potential on pacemaker activity of canine Purkinje fibers in relation to parasystole. *Circ. Res.* 39: 801–8. [6.1, 6.4, 7.1, 7.5, 9.3]

————. 1979. A biologic model for parasystole. *Am. J. Cardiol.* 43: 761–72. [6.1, 6.4, 7.1, 7.5, 9.3]

Jalife, J., Antzelevitch, C., and Moe, G. K. 1982. The case for modulated parasystole. *PACE* 5: 911–26. [6.4]

Jensen, J. H., Christiansen, P. L., and Scott, A. C. 1984. Chaos in the Beeler-Reuter system for the action potential of ventricular myocardial fibres. *Physica* 13D: 269–77. [7.4]

Jensen, J. H., Christiansen, P. L., Scott, A. C., and Skovgaard, O. 1983. Chaos in nerve. Proceedings of the Iasted Symposium, Copenhagen, Denmark, ACI, 2, 15/6–15/9. [7.4]

Jensen, M. H., Bak, P., and Bohr, T. 1984. Transition to chaos by interaction of resonances in dissipative systems. I: Circle maps. *Phys. Rev.* 30A: 1960–69. [A.2]

Josephson, M. E., Buxton, A. E., Marchlinski, F. E., Doherty, J. V., Cassidy, D. M., Kienzle, M. G., Vassalo, J. A., Miller, J. M., Almendral, J., and Grogan, W. 1985. Sustained ventricular tachycardia in coronary artery disease—Evidence for a reentry mechanism. In *Cardiac Electrophysiology and Arrhythmias*, ed. D. P. Zipes and J. Jalife. pp. 409–18. Orlando: Grune and Stratton. [8.2]

Joyner, R. W., Picone, J., Veenstra, R., and Rawling, D. 1983. Propagation through electrically coupled cells. Effects of regional changes in membrane properties. *Circ. Res.* 53: 526–34. [8.1]

Kaczmarek, L. K., and Babloyantz, A. 1977. Spatio-temporal patterns in epileptic seizures. *Biol. Cybern.* 26: 199–208. [4.6]

Kaplan, J. L., and Yorke, J. A. 1979. Chaotic behavior of multidimensional difference equation. In *Functional Difference Equations and Approximations of Fixed Points*, H. O. Peitgen and H. O. Walther, pp. 228–37. New York: Springer-Verlag. [3.4]

Katz, L. N. 1946. *Electrocardiology.* 2d ed. Philadelphia: Lea and Febiger. [9.1]

Kauffman, S., and Wille, J. J. 1975. The mitotic oscillator in *Physarum polycephalum. J. Theor. Biol.* 55: 47–93. [5.1]

Kawato, M. 1981. Transient and steady phase response curves of limit cycle oscillators. *J. Math. Biol.* 12: 13–30. [6.3]

Kawato, M., and Suzuki, R. 1978. Biological oscillators can be stopped. Topological study of a phase response curve. *Biol. Cybern.* 30: 241–48. [6.3]

Kazarinoff, N. D., and Van den Driessche, P. 1979. Control of oscillations in hematopoesis. *Science* 203: 1348–49. [5.2]

Keener, J. P. 1980. Chaotic behavior in piecewise continuous difference equations. *Trans. Am. Math. Soc.* 261: 589–604. [7.3, 8.1]

———. 1981. On cardiac arrhythmias: AV conduction block. *J. Math. Biol.* 12: 215–25. [7.3, 8.1]

———. 1986. Spiral waves in excitable media. In *Nonlinear Oscillations in Biology and Chemistry*, ed. H. G. Othmer, pp. 115–27. Berlin: Springer-Verlag. [8.3]

Keener, J. P., and Glass, L. 1984. Global bifurcations of a periodically forced oscillator. *J. Math. Biol.* 21: 175–90. [2.3, 7.4]

Keener, J. P., Hoppensteadt, F. C., and Rinzel, J. 1981. Integrate and fire models of nerve membranes response to oscillatory inputs. *SIAM J. Appl. Math.* 41: 503–17. [7.3]

Keller, E. F. 1967. A mathematical description of biological clocks. *Currents in Mod. Biol.* 1: 279–84. [7.4]

Khoo, M.C.K., Kronauer, R. E., Strohl, K. P., and Slutsky, A. S. 1982. Factors inducing periodic breathing in humans: A general model. *J. Appl. Physiol.* 53: 644–59. [4.5]

Killman, S. A., Cronkite, E. P., Robertson, J. S., Fliedner, T. M., and Bond, V. P. 1963. Estimation of phases of the life cycle of leukemic cells from labeling in human beings *in vivo* with tritiated thymidine. *Lab. Invest.* 12: 671–84. [4.7]

Kiloh, L. G., McComas, A. J., Osselton, J. W., and Upton, A.R.M. 1981. *Clinical Electroencephalography.* London: Butterworths. [1.1]

King, R., Barchas, J. D., and Huberman, B. A. 1984. Chaotic behavior in dopamine neurodynamics. *Proc. Nat. Acad. Sci. USA* 81: 1244–47. [4.4]

King-Smith, E. A., and Morley, A. 1970. Computer simulation of granulopoiesis: Normal and impaired granulopoiesis. *Blood* 36: 254–62. [9.2]

Kitney, R. I., and Rompelman, O., eds. 1980. *The Study of Heart Rate Variability*. Oxford: Clarendon Press [3.3, 9.1]

Kling, V., and Szekely, G. 1968. Simulation of rhythmic nervous activities. I: Function of networks with cyclic inhibitions. *Kybernetik* 5: 89–103. [4.4, 5.2]

Knight, B. W. 1972. Dynamics of encoding in a population of neurons. *J. Gen. Physiol.* 59: 734–66. [7.3]

Knobil, E. 1974. On the control of gonadotropin secretion in Rhesus monkey. *Rec. Prog. Horm. Res.* 30: 1–46. [1.1]

Knowles, W. D., Traub, R. D., Wong, R.K.S., and Miles, R. 1985. Properties of neural networks—Experimentation and modeling of the epileptic hippocampal slice. *Trends Neuroscience* 8: 73–79. [4.6]

Knox, C. K. 1973. Characteristics of inflation and deflation reflexes during expiration in the cat. *J. Neurophysiol.* 36: 284–95. [6.1]

Kobayashi, M., and Musha, T. 1982. 1/f fluctuation of heartbeat period. *IEEE Trans. Biomed. Eng.* BME-29: 456–57 [3.3, 9.1]

Kopell, N. 1986. Coupled oscillators and locomotion by fish. In *Nonlinear Oscillations in Chemistry and Biology*, ed. H. G. Othmer, pp. 160–74. Berlin: Springer-Verlag. [8.1]

Kopell, N., and Ermentrout, G. B. 1983. Coupled oscillators in mammalian small intestine. In *Oscillations in Mathematical Biology*. ed. J.P.E. Hodgson, pp. 24–36. Berlin: Springer-Verlag. [8.1]

Kopell, N., and Howard, L. N. 1973. Plane wave solutions to reaction-diffusion equations. *Studies in Appl. Math.* 52: 291–328. [2.3]

———. 1981. Target patterns and spiral solutions to reaction-diffusion equations with more than one space dimension. *Adv. Appl. Math.* 2: 417–49. [8.3]

Koslow, S. H., Mandell, A. J., and Schlesinger, M. F. 1987. *Perspectives in Biological Dynamics and Theoretical Medicine*. Ann. N.Y. Acad. Sci., vol. 504. [1.1]

Kostelich, E. J., and Swinney, H. L. 1987. Practical considerations in estimating dimension from time series data. In *Chaos and Related Nonlinear Phenomena*, I. Procaccia and M. Shapiro. New York: Plenum. [3.4, 4.6]

Krinskii, V. I. 1968. Fibrillation in excitable media. *Systems Theory Research (Prob. Kyb.)* 20: 46–65. [8.2, 8.5]

Kripke, D. F. 1983. Phase advance theories for affective illness. In *Circadian Rhythms in Psychiatry*, ed. T. A. Wehr and F. K. Goodwin, pp. 41–69. Pacific Grove, Calif.: The Boxwood Press. [7.5]

Kronecker, H. 1886. Ueber Störungen der Coordination des Herzkammerschlages. Zeitschr. f. Biol. 34: 529–603. [8.5]

Kronenberg, F., Cote, L. J., Linkie, D. M., Dyrenfurth, I., and Downey, J. A. 1984. Menopausal hot flashes: Thermoregulatory, cardiovascular, and circulating catecholamine and LH changes. *Maturitas* 6: 31–43. [5.3]

Lambertsen, C. J. 1974. Abnormal types of respiration. In *Handbook of Medical Physiology*, 13th ed., ed. V. Mountcastle, pp. 1522–37. St. Louis: The C. V. Mosby Co. [4.5]

Langford, W. F. 1977. Numerical solution of bifurcation problems for ordinary differential equations. *Numer. Math.* 28: 171–90. [5.2]

Lasota, A. 1977. Ergodic problems in biology. *Astérisque* 50: 239–50. [4.6, 9.2]

Lasota, A., and Mackey, M. C. 1980. The extinction of slowly evolving dynamical systems. *J. Math. Biol.* 10: 333–45. [3.2, 9.4]

——. 1985. *Probabilistic Properties of Deterministic Systems.* Cambridge: Cambridge University Press. [3.1, 3.2, 3.3, 7.3, A.1, A.2]

Lasota, A., Mackey, M. C., and Wazewska-Czyzewska, M. 1981. Minimizing therapeutically induced anemia. *J. Math. Biol.* 13: 149–58. [9.4]

Leao, A.A.P. 1944. Spreading depression of activity in the cerebral cortex. *J. Neurophysiol.* 7: 359–90. [8.5]

Le Berre, M., Ressayre, E., Tallet, A., Gibbs, H. M., Kaplan, D. L., and Rose, M. H. 1987. Conjecture on the dimensions of chaotic attractors of delayed-feedback dynamical systems. *Phys. Rev. A.* 35: 4020–2. [4.6]

Lebrun, P., and Atwater, I. 1985. Chaotic and irregular bursting of electrical activity in mouse pancreatic β-cells. *Biophys. J.* 48: 529–31. [4.1]

Lee, R. G., and Stein, R. B. 1981. Resetting of tremor by mechanical perturbations: A comparison of essential tremor and parkinsonian tremor. *Ann. Neurol.* 10: 523–31. [6.4]

Levi, M. 1981. Qualitative analysis of the periodically forced relaxation oscillations. *Mem. Am. Math. Soc.*, no. 244. [7.2]

Levinson, N. 1949. A second order differential equation with singular solutions. *Annals of Mathematics* 50: 127–53. [7.2]

Levy, M. N., Iano, T., and Zieske, H. 1972. Effects of repetitive bursts of vagal activity on heart rate. *Circ. Res.* 30: 186–95. [7.1]

Levy, M. N., Martin, P. J., Edelstein, J., and Goldberg, L. B. 1974a. The AV nodal Wenckebach phenomenon as a positive feedback mechanism. *Prog. Cardiovasc. Dis.* 16: 601–13. [8.1]

Levy, M. N., Martin, P. J., Zieske, H., and Adler, D. 1974b. Role of positive feedback in the atrioventricular nodal Wenckebach phenomenon. *Circ. Res.* 34: 697–710. [8.1]

Lewis, T. 1920. Observations upon flutter and fibrillation, pt. 4: Impure flutter; theory of circus movement. *Heart* 7: 293–345. [8.2, 8.3]

————. 1925. *The Mechanism and Graphic Registration of the Heart.* London: Shaw and Sons. [8.2, 8.3]

Lewis, T., and Master, A. M. 1925. Observations upon conduction in the mammalian heart. A-V conduction. *Heart* 12: 209–70. [8.1]

Lewy, A. J., Sack, R. L., and Singer, C. M. 1985. Treating phase typed chronobiologic sleep and mood disorders using appropriately timed bright artifical light. *Psychopharmacol. Bull.* 21: 368–72. [7.5, 9.4]

Li, T.-Y., and Yorke, J. A. 1975. Period three implies chaos. *Amer. Math. Monthly* 82: 985–92. [2.5]

Lippold, O.C.J. 1970. Oscillation in the stretch reflex arc and the origin of the rhythmical 8–12 c/s component of physiological tremor. *J. Physiol. (Lond.)* 206: 359–82. [4.5]

Loidl, P., and Sachsenmaier, W. 1982. Control of mitotic synchrony in *Physarum polycephalum*: Phase shifting by fusion of heterophasic plasmodia contradicts a limit cycle oscillator model. *Eur. J. Cell Biol.* 28: 175–79. [5.1]

Longobardo, G. S., Cherniack, N. S., and Fishman, A. P. 1966. Cheyne-Stokes breathing produced by a model of the human respiratory system. *J. Appl. Physiol.* 21: 1839–46. [4.5]

Longtin, A., and Milton, J. G. 1988. Complex oscillations in the human pupil light reflex with 'mixed' and delayed feedback. *Math. Biosci.* In Press. [4.4]

Lorenz, E. N. 1963. Deterministic nonperiodic flow. *J. Atmos. Sci.* 20: 282–93. [2.5, A.1]

Lown, B. 1979. Sudden cardiac death. The major challenge confronting contemporary cardiology. *Am. J. Cardiol.* 43: 313–28. [8.5]

Lund, J. E., Padgett, G. A., and Oh, R. L. 1967. Cyclic neutropenia in grey collie dogs. *Blood* 29: 452–6.1, [9.3]

Lund, J. P., and Dellow, P. G. 1973. Rhythmical masticatory activity of hypoglossal motoneurons responding to an oral stimulus. *Exp. Neurol.* 40: 243–46. [5.3]

Lund, J. P., Rossignol, S., and Murakami, T. 1981. Interactions between the jaw opening reflex and mastication. *Can. J. Physiol. Pharmacol.* 59: 683–90. [6.4]

McAllister, R. E., Noble, D., and Tsien, R. W. 1975. Reconstruction of the electrical activity of cardiac Purkinje fibers. *J. Physiol. (Lond.)* 251: 1–59. [2.1, 4.1]

McClellan, A. D., and Grillner, S. 1984. Activation of 'fictive swimming' by electrical microstimulation of brainstem locomotor regions in an *in vitro* preparation of the lamprey central nervous system. *Brain Res.* 300: 357–61. [5.2]

Mackay, R. S., and Tresser, C. 1986. Transition to topological chaos for circle maps. *Physica* 19D: 206–37. [A.2]

Mackey, M. C. 1978. A unified hypothesis for the origin of aplastic anemia and periodic hematopoiesis. *Blood* 51: 941–56. [9.4, A.1]

———. 1979a. Dynamic haematological disorders of stem cell origin. In *Biophysical and Biochemical Information Transfer in Recognition*, ed. J. Vassilova-Popova and E. V. Jensen, pp. 373–409. New York: Plenum. [4.6, A.1]

———. 1979b. Periodic auto-immune hemolytic anemia: An induced dynamical disease. *Bull. Math. Biol.* 41: 829–34. [4.5, A.1]

———. 1985. Deterministic cell cycle models with transition probability like properties. In *Temporal Order*, ed. L. Rensing and N. I. Jaeger, pp. 315–20. Berlin: Springer-Verlag. [3.2]

Mackey, M. C., and Dörmer, P. 1981. Enigmatic hemopoiesis. In *Biomathematics and Cell Kinetics*, ed. M. Rotenberg, pp. 87–103. Amsterdam: Elsevier/North-Holland Biomedical Press. [9.4]

———. 1982. Continuous maturation of proliferating erythroid precursors. *Cell Tissue Kinet.* 15: 381–92. [9.4]

Mackey, M. C., and Glass, L. 1977. Oscillation and chaos in physiological control systems. *Science* 197: 287–89. [1.2, 4.5, 4.6, 9.1, A.1]

Mackey, M. C., and an der Heiden, U. 1982. Dynamical diseases and bifurcations: Understanding functional disorders in physiological systems. *Funkt. Biol. Med.* 1: 156–64. [9.1]

———. 1984. The dynamics of recurrent inhibition. *J. Math. Biol.* 19: 211–25. [4.6, A.1]

Mackey, M. C., and Milton, J. G. 1987. Dynamical diseases. *Ann. N.Y. Acad. Sci.* 504: 16–32. [9.1]

Mackey, M. C., Santavy, M., and Selepova, P. 1986. A mitotic oscillator model for the cell cycle with a strange attractor. In *Nonlinear Oscillations in Biology and Chemistry*, ed. H. G. Othmer, pp. 34–45. Berlin: Springer-Verlag. [3.2]

Madore, B. F., and Freedman, W. L. 1983. Computer simulations of the Belousov-Zhabotinsky reaction. *Science* 222: 615–16. [8.3]

Mandelbrot, B. B. 1977. *Fractals: Form, Chance and Dimension.* San Francisco: W. H. Freeman. [3.4, 8.1]

———. 1982. *The Fractal Geometry of Nature.* San Francisco: W. H. Freeman. [3.4, 8.1]

Marik, J., and Hulka, J. 1978. Luteinized unruptured follicle syndrome: A subtle cause of infertility. *Fertil. Steril.* 29: 270–74. [5.1]

Marriott, H.J.L., and Conover, M. H. 1983. *Advanced Concepts in Cardiac Arrhythmias.* St. Louis: C. V. Mosby. [8.1, 8.2, 9.1]

Marsden, J. E., and McCracken, M. 1976. *The Hopf Bifurcation and Its Applications.* New York: Springer-Verlag. [5.2, 5.3, A.1]

Martiel, J. L., and Goldbeter, A. 1985. Autonomous chaotic behavior of the slime mold *Dictyostelium discoideum* predicted by a model for cyclic AMP signalling. *Nature* 313: 590–92. [4.1]

Martins-Ferreira, H., de Oliveira Castro, G., Struchiner, C. J., and Rodrigues, P. S. 1974. Circling spreading depression in isolated chick retina. *J. Neurophysiol.* 37: 773–84. [8.5]

Mates, J.W.B., and Horowitz, J. M. 1976. Instability in a hippocampal neural network. *Comp. Prog. Biomed.* 6: 74–84. [4.6]

May, R. M. 1973. *Stability and Complexity in Model Ecosystems.* Princeton: Princeton University Press. [A.1]

————, 1976. Simple mathematical models with very complicated dynamics. *Nature* 261: 459–67. [2.5]

————. 1980. Nonlinear phenomena in ecology and epidemiology. *Ann. N.Y. Acad. Sci.* 357: 267–81. [4.6, A.3]

May, R. M., and Oster, G. F. 1976. Bifurcations and dynamic complexity in simple ecological models. *Amer. Natur.* 110: 573–99. [2.5]

————. 1980. Period doubling and the onset of turbulence: An analytic estimate of the Feigenbaum ratio. *Phys. Lett.* 78A: 1–3. [2.5]

Mayer, A. G. 1908. Rhythmical pulsation in scyphomedusae. *Papers of the Tortugas Lab of the Carnegie Inst. of Wash.* 6: 25–54. [8.2]

Mayer-Kress, G., ed. 1986. *Dimensions and Entropies in Chaotic Systems: Quantification of Complex Dynamics.* Berlin: Springer-Verlag. [3.3, 3.4]

Merton, P. A., Morton, H. B., and Rashbass, C. 1967. Visual feedback in hand tremor. *Nature* 216: 583–84. [4.5]

Metropolis, N., Stein, M. L., and Stein, P. R. 1973. On finite limit sets for transformations on the unit interval. *J. Comb. Theory* 15: 25–44. [2.5]

Mikhailov, A. S., and Krinskii, V. I. 1983. Rotating spiral waves in excitable media: The analytic results. *Physica* 9D: 346–71. [8.3]

Milhorn, H. T. 1966. *The Application of Control Theory to Physiological Systems.* Philadelphia: W. B. Saunders. [4.5, 9.2]

Millhorn, D. E., Eldridge, F. L., Kiley, J. P., and Waldrop, T. G. 1984. Prolonged inhibition of respiration following acute hypoxia in glomectomized cats. *Respir. Physiol.* 57: 331–40. [5.2]

Mines, G. R. 1913. On dynamic equilibrium in the heart. *J. Physiol. (Lond.)* 46: 349–82. [8.1, 8.2, 8.3]

————. 1914. On circulating excitation on heart muscles and their possible relation to tachycardia and fibrillation. *Trans. R. Soc. Can.* 4: 43–53. [8.2, 8.3, 8.5]

Miura, R. M., and Plant, R. E. 1981. Rotating waves in models of excitable media. In *Differential Equations and Applications in Ecology, Epidemics and Population Problems*, ed. S. N. Busenberg and K. L. Cooke, pp. 247–57. New York: Academic Press. [8.3]

Mobitz, W. 1924. Über die unvollständige Störung der Erregungsüberleitung zwischen Vorhof und Kammer des menschlichen Herzens. *Zeit. f. d. ges. Exp. Med.* 41: 180–237. [8.1, 9.2]

Moe, G. K., and Abildskov, J. A. 1959. Atrial fibrillation as a self-sustaining arrhythmia independent of focal discharge. *Am. Heart J.* 58: 59–70. [8.5]

Moe, G. K., Rheinboldt, W. C., and Abildskov, J. A. 1964. A computer model of atrial fibrillation. *Am. Heart J.* 67: 200–20. [8.3, 8.5]

Moe, G. K., Jalife, J., Mueller, W. J., and Moe, B. 1977. A mathematical model of parasystole and its application to clinical arrhythmias. *Circulation* 56: 968–79. [6.4, 7.4, 7.5]

Molnar, G. D., Taylor, W. F., and Langworthy, A. L. 1972. Plasma immunoreactive insulin patterns in insulin-treated diabetics. *Mayo Clin. Proc.* 47: 709–19. [1.1]

Moore-Ede, M. C., and Czeisler, C. A. eds. 1984. *Mathematical Models of the Circadian Sleep-Wake Cycle*. New York: Raven Press. [7.5]

Muller, S. L., Plesser, T., and Hess, B. 1985. The structure of the core of the spiral wave in the Belousov-Zhabotinskii reaction. *Science* 230: 661–63. [8.3]

Murray, J. D. 1988. *Mathematical Biology*. Berlin: Springer-Verlag. [A.1, A.2]

Nagumo, J., Suzuki, R., and Sato, S. 1963. *Electrochemical Active Network*. Notes of professional group on nonlinear theory of IECE (Japan). [8.3]

Nau, G. J., Aldariz, A. E., Acunzo, R. S., Halpern, M. S., Davidenko, J. M., Elizari, M. V., and Rosenbaum, M. B. 1982. Modulation of parasystolic activity by nonparasystolic beats. *Circulation* 66: 462–69. [6.4]

Nikolic, G., Bishop, R. L., and Singh, J. B. 1982. Sudden death recorded during Holter monitoring. *Circulation* 66: 218–25. [8.5, 9.1]

Noble, D. 1983. Ionic mechanisms of rhythmic firing. *Symp. Soc. Exp. Biol.* 37: 1–28. [2.1, 4.1]

————. 1984. The surprising heart: A review of recent progress in cardiac electrophysiology. *J. Physiol. (Lond.)* 353: 1–50. [2.1, 4.1]

Nolasco, J. B., and Dahlen, R. W. 1968. A graphic method for the study of alternation in cardiac action potentials. *J. Appl. Physiol.* 25: 191–96. [8.1]

Ogawa, M., Fried, J., Sakai, Y., Strife, A., and Clarkson, B. D. 1970. Studies of cellular proliferation in human leukemia. VI: The proliferative activity, generation time, and emergence time of neutrophilic granulocytes in chronic granulocytic leukemia. *Cancer* 25: 1031–49. [4.6]

Olsen, L. F., and Degn, H. 1985. Chaos in biological systems. *Quart. Rev. Biophys.* 18: 165–225. [1.1]

Ostlund, S., Rand, D., Sethna, J., and Siggia, E. 1983. Universal properties of the transition from quasi-periodicity to chaos in dissipative systems. *Physica* 8D: 303–42. [A.2]

Othmer, H. G., ed. 1986. *Nonlinear Oscillations in Biology and Chemistry. Lecture Notes in Biomathematics*, vol. 66 Berlin: Springer-Verlag. [1.1]

Pack, A. I., and Millman, R. P. 1986. Changes in control of ventilation, awake and asleep in the elderly. *J. Am. Soc. Ger.* 34: 533–44. [4.5]

Patton, R. J., and Linkens, D. A. 1978. Hodgkin-Huxley type electronic modelling of gastrointestinal electrical activity. *Med. and Biol. Eng. and Computing* 16: 195–202. [8.1]

Pavlidis, T. 1973. *Biological Oscillators: Their Mathematical Analysis* New York: Academic Press. [1.2, 6.3, 7.1, 7.4]

Paydarfar, D., and Eldridge, F. L. 1987. Phase resetting and dysrhythmic responses of the respiratory oscillator. *Am. J. Physiol.* 252 (*Regulatory Integrative Comp. Physiol.* 21): R55–62. [5.4, 6.4]

Paydarfar, D., Eldridge, F. L., and Kiley, J. P. 1986. Resetting of the mammalian respiratory rhythm: Existence of a phase singularity. *Am. J. Physiol.* 250 (*Regulatory Integrative Comp. Physiol.* 19): R721–27. [5.4, 6.3]

Peitgen, H.-O., and Richter, P. H. 1986. *The Beauty of Fractals*. Berlin: Springer-Verlag. [3.4]

Perez, R., and Glass, L. 1982. Bistability, period doubling bifurcations and chaos in a periodically forced oscillator. *Phys. Lett.* 90A: 441–43. [A.2]

Perez, J. F., Malta, C. P., and Coutinho, F.A.B. 1978. Quantitative analysis of oscillations in isolated populations of flies. *J. Theor. Biol.* 71: 505–14. [4.6]

Perkel, D. H., and Mulloney, B. 1974. Motor production in reciprocally inhibiting neurons exhibiting post-inhibitory rebound. *Science* 185: 181–83. [4.3]

Perkel, D. H., Schulman, J. H., Bullock, T. H., Moore, G. P., and

Segundo, J. P. 1964. Pacemaker neurons: Effects of regularly spaced synaptic input. *Science* 145: 61–63. [7.1, 7.4]

Peters, H. 1980. Comportement chaotique d'une équation différentielle retardée. *C. R. Acad. Sci. Paris* 290A: 1119–22. [4.6]

Petersen, I., and Stener, I. 1970. An electromyographical study of the striated urethral sphincter, the striated anal sphincter. and the levator ani muscle during ejaculation. *Electromyography* 10: 23–44. [5.3]

Petrillo, G. A., and Glass, L. 1984. A theory for phase locking respiration in cats to a mechanical ventilator. *Am. J. Physiol.* 246 (*Regulatory Integrative Comp. Physiol.* 15): R311–20. [4.4, 5.2, 6.2, 7.1, 7.3]

Petrillo, G. A., Glass, L., and Trippenbach, T. 1983. Phase locking of the respiratory rhythm in cats to a mechanical ventilator. *Can. J. Physiol. Pharmacol.* 61: 599–607. [7.1]

Pham Dinh, T., Demongeot, J., Baconnier, P., and Benchetrit, G. 1983. Simulation of a biological oscillator: The respiratory system. *J. Theor. Biol.* 103: 113–32. [6.2]

Pinsker, H. M. 1977. Aplysia bursting neurons as endogenous oscillators. II: Synchronization and entrainment by pulsed inhibitory synaptic input. *J. Neurophysiol.* 40: 544–52. [7.1, 7.4]

Pittendrigh, C. S. 1965. On the mechanism of entrainment of a circadian rhythm by light cycles. In *Circadian Clocks*, ed. J. Aschoff, pp. 277–97. Amsterdam: North Holland. [7.1, 7.4]

Plant, R. E. 1982. The analysis of models for excitable membranes: An introduction. *Lectures on Mathematics in the Life Sciences* 15: 27–54. [5.2, 5.3]

Poincaré, H. 1881. Mémoire sur les courbes définies par une équation différentielle. *J. de Math.*, 3d ser., 7: 375–422. [2.3]

———. 1882. Mémoire sur les courbes définies par une équation différentielle. *J. de Math*, 3d ser., 8: 251–96. [2.3]

———. 1885. Sur les courbes définies par les équations différentielles. *J. Math. Pures et Appliq*, 4th ser., 1: 167–244. [2.3, 7.2]

———. 1954. *Ouevres I*. Paris: Gauthier-Villar. [2.3, 7.2]

Polosa, C. 1984. Rhythms in the activity of the autonomic nervous system: Their role in the generation of systemic arterial pressure waves. In *Mechanisms of Blood Pressure Waves*, K. Miyakawa, H. P. Koepchen, and C. Polosa, pp. 27–41. Tokyo: Japan Scientific Societies Press. [1.1]

Prosser, C. L., Smith, C. E., and Melton, C. E. 1955. Conduction of action potentials in the ureter of the rat. *Am. J. Physiol.* 181: 651–60. [8.1]

Publicover, N. G., and Sanders, K. M. 1986. Effects of frequency on the wave form of propagated slow waves in canine gastric antral muscle. *J. Physiol.* 371: 179–89. [8.1]

Quesenberry, P., and Levitt, L. 1979. Hematopoietic stem cells. *New Eng. J. Med.* 301: 755–60, 819–23. [4.6]

Rand, R. H., Cohen, A. H., and Holmes, P. J. 1988. Systems of coupled oscillators as models of CPGs. In *Neural Control of Rhythmic Movements in Vertebrates*, ed. A. H. Cohen, S. Rossignol, and S. Grillner. pp. 333–67. New York: Wiley. [8.1]

Rapp, P. E., and Berridge, M. J. 1977. Oscillations in calcium-cyclic AMP control loops form the basis of pacemaker activity and other high frequency biological rhythms. *J. Theor. Biol.* 66: 497–525. [5.1]

Rapp, P. E., Zimmerman, I. D., Albano, A. M., de Guzman, G. C., Greenbaun, M. N., and Bashore, T. R. 1986. Experimental studies of chaotic neural behavior: Cellular activity and electroencephalographic signals. In *Nonlinear Oscillations in Biology and Chemistry*, ed. H. G. Othmer, pp. 175–205. Berlin: Springer-Verlag. [3.3, 3.4]

Reid, J.V.O. 1969. The cardiac pacemaker: Effects of regularly spaced nervous input. *Am. Heart J.* 78: 58–64. [7.1]

Reimann, H. A. 1963. *Periodic Diseases*. Philadelphia: F. A. Davis. [9.1]
———. 1974, Clinical importance of biorhythms longer than the circadian. In *Chronobiology*, ed. L. E. Schering, F. Halberg, and J. E. Pauly, pp. 304–305. Tokyo: Igaku Shoin. [9.1]

Remmers, J. E. 1976. Analysis of ventilatory response. *Chest* 70: Suppl. 1. 134–37. [6.2]

Rensing, L., an der Heiden, U., and Mackey, M. C., eds. 1987. *Temporal Disorder in Human Oscillatory Systems*. Berlin: Springer-Verlag. [1.1, 9.1]

Rescigno, A., Stein, R. B., Purple, R. L., and Poppele, R. E. 1970. A neuronal model for the discharge patterns produced by cyclic inputs. *Bull. Math. Biophys.* 32: 337–53. [7.3]

Reshodko, L. V., and Bures, J. 1975. Computer simulation of reverberating spreading depression in a network of cell automata. *Biol. Cyb.* 18: 181–89. [8.3]

Reynolds, S.R.M. 1965. *Physiology of the Uterus*. New York: Hafner. [5.2]

Richter, C. P. 1965. *Biological Clocks in Medicine and Psychiatry*. Springfield; Ill.: C. C. Thomas. [9.1]

Richter, D. W., and Ballantyne, D. 1983. A three phase theory about the basic respiratory pattern generator. In *Central Neurone En-*

vironment, ed. M. E. Schlafke, H. P. Koechen, and W. R. See, pp. 165–74. Berlin: Springer-Verlag. [4.4]

Rinzel, J. 1980. Impulse propagation in excitable systems. In *Dynamics and Modelling of Reactive Systems*, ed. W. E. Stewart, W. H. Ray, and C. C. Conley, pp. 259–91. New York: Academic Press. [8.1]

———. 1981. Models in neurobiology. In *Lectures in Applied Mathematics*, vol. 19: *Mathematical Aspects of Physiology*, pp. 281–97. Providence: American Mathematical Society. [8.1]

Rinzel, J., and Maginu, K., 1984. Kinematic analysis of wave pattern formation in excitable media. In *Non-Equilibrium Dynamics in Chemical Systems*, ed. C. Vidal and A. Pacault, pp. 107–13. Berlin: Springer-Verlag. [8.1]

Rinzel, J., and Miller, R. N. 1980. Numerical solutions of stable and unstable periodic solutions to the Hodgkin-Huxley equations. *Math. Biosci.* 49: 27–59. [5.3, 8.1]

Ritzenberg, A. L., Smith, J. M., Grumbach, M. P., and Cohen, R. J. 1984. Precursor to fibrillation in cardiac computer model. In *Computers in Cardiology*, 171–74. Long Beach, Calif.: IEEE Computer Society. [8.5]

Robinson, G. C. 1913. The influence of the vagus nerves on the faradized auricles in the dog's heart. *J. Exp. Med.* 17: 429–43. [8.5]

Rodieck, R. W., Kiang, N. Y.-S., and Gerstein, G. 1962. Some quantitative methods for the study of spontaneous activity of single neurons. *Biophys. J.* 2: 351–68. [3.1]

Rohlicek, C. V., and Polosa, C. 1983. Mediation of pressor responses to cerebral ischemia by superficial ventral medullary areas. *Am. J. Physiol.* 245 (*Heart Circ. Physiol.* 14): H962–68. [5.2]

Rössler, O. E. 1979. Continuous chaos—four prototype equations. *Ann. N.Y. Acad. Sci.* 316: 376–92. [A.1]

Ruelle, D., and Takens, F. 1971. On the nature of turbulence. *Commun. Math. Phys.* 20: 167–92; 23: 343–44. [3.4]

Sagawa, K., Carrier, O., and Guyton, A. C. 1962. Elicitation of theoretically predicted feedback oscillation in arterial pressure. *Am. J. Physiol.* 203: 141–46. [4.5]

Sakmann, B., Noma, A., and Trautwein, W., 1983. Acetylcholine activation of single muscarinic K^+ channels in isolated pacemaker cells of the mammalian heart. *Nature* 303: 250–53. [1.1, 3.1]

Salmoiraghi, G. C., and Burns, B. D. 1960. Notes on mechanism of rhythmic respiration. *J. Neurophysiol.* 23: 14–26. [4.3]

Sarna, S. K. 1985. Cyclic motor activity: Migrating motor complex. *Gastroenterology* 89: 894–913. [8.1]

Sarna, S. K., Daniel, E. E., and Kingma, Y. J. 1971. Simulation of slow wave electrical activity of small intestine. *Amer. J. Physiol.* 221: 161–75. [8.1]

Satterlie, R. A. 1985. Reciprocal inhibition and postinhibitory rebound produce reverberation in a locomotor pattern generator. *Science* 229: 402–404. [4.3]

Saupe, D. 1982. Beschleunigte PL-Kontinuitatsmethoden und periodische Lösungen parameterisierter Differentialgleichungen mit Zeitverzögerung. Doctoral dissertation, University of Bremen, FRG. [4.6]

Schell, M., Fraser, S., and Kapral, R. 1983. Subharmonic bifurcation in the sine map: An infinite hierarchy of cusp bistabilities. *Phys. Rev. A* 28: 373–78.

Schulman, H., Duvivier, R., and Blattner, P. 1983. The uterine contractility index. *Am. J. Obstet. Gynecol.* 145: 1049–58. [5.2]

Schuster, H. G. 1984. *Deterministic Chaos.* Weinheim, FRG: Physik-Verlag. [A.1, A.2]

Scott, S. W. 1979. Stimulation simulations of young yet cultured beating hearts. Ph.D. dissertation, SUNY at Buffalo. [2.3, 7.1, 7.4]

Segel, L. A. 1984. *Modelling Dynamic Phenomena in Molecular and Cellular Biology.* Cambridge: Cambridge University Press. [A.1, A.2]

Segundo, J. P., and Kohn, A. F. 1981. A model of excitatory synaptic interactions between pacemakers: Its reality, its generality and principles involved. *Biol. Cybern.* 40: 113–26. [7.4]

Selfridge, O. 1948. Studies on flutter and fibrillation. V: Some notes on the theory of flutter. *Arch. Inst. Cardiol. Mex.* 18: 177–87. [8.3]

Selverston, A. I., Miller, J. P., and Wadepuhl, M. 1983. Cooperative mechanisms for the production of rhythmic movements. In *Neural Origin of Rhythmic Movements*, ed. A. Roberts and B. Roberts, pp. 55–87. Soc. Exp. Biol. Symposium 37. [4.3]

Shibata, M., and Bures, J. 1972. Reverberation of cortical spreading depression along closed-loop pathways in rat cerebral cortex. *J. Neurophysiol.* 35: 381–88. [8.5]

Shimkin, M. B., Mettier, S. R., and Bierman, H. R. 1950. Myelocytic leukemia: An analysis of incidence, distribution and fatality, 1910–1948. *Ann. Intern. Med.* 35: 194–212. [9.4]

Shrier, A., Dubarsky, H., Rosengarten, M., Guevara, M. R., Nattel, S., and Glass, L. 1987. Prediction of atrioventricular conduction rhythms in humans using the atrioventricular nodal recovery curve. *Circulation.* 76: 1196–1205. [8.1]

Shymko, R. M., and Glass, L. 1974. Spatial switching in chemical reactions with heterogeneous catalysis. *J. Chem. Phys.* 60: 835–41. [A.1]

Siegel, G., Ebeling, B. J., Hofer, H. W., Nolte, J., Roedel, H., and Klubendorf, D. 1984. Vascular smooth muscle rhythmicity. In *Mechanisms of Blood Pressure Waves*, ed. K. Miyakawa, H. P. Koepchen, and C. Polosa, pp. 319–40. Tokyo: Japan Scientific Societies Press. [1.1]

Smale, S. 1967. Differentiable dynamical systems. *Bull Am. Math. Soc.* 73: 748–817. [3.3]

Smith, J. A., and Martin, L. 1973. Do cells cycle? *Proc. Natl. Acad. Sci. USA* 70: 1263–67. [3.2]

Smith, J. M., and Cohen, R. J. 1984. Simple finite-element models account for wide range of cardiac dysrhythmias. *Proc. Nat. Acad. Sci. USA* 81: 233–37. [8.3, 8.5]

Stark, L. W. 1968. *Neurological Control Systems: Studies in Bioengineering*. New York: Plenum. [4.5, 9.2]

———. 1984. The pupil as a paradigm for neurological control systems. *IEEE Trans. on Biomed. Eng.* (BME-31): 919–24. [4.5]

Stein, P.S.G. 1977. Application of the mathematics of coupled oscillator systems to the analysis of the neural control of locomotion. *Federation Proc.* 36: 2056–59. [7.5]

Stein, R. B., Lee, R. G., and Nichols, T. R. 1978. Modifications of ongoing tremors and locomotion by sensory feedback. *Electroencephelogr. Clin. Neurophysiol.* (Suppl) 34: 511–19. [6.4]

Strogatz, S. H. 1986. *The Mathematical Structure of the Human Sleep-Wake Cycle. Lecture Notes in Biomathematics*, vol. 69. Berlin: Springer-Verlag. [7.3]

Swinney, H. L. 1983. Observations of order and chaos in nonlinear dynamics. *Physica* 7D: 3–15. [3.3]

Szekely, G. 1965. Logical network for controlling limb movements in Urodela. *Acta Physiol. Hung* 27: 285–89. [4.4, 5.2]

Thom, R. 1970. Topological models in biology. In *Towards a Theoretical Biology*. 3: *Drafts*, ed. C. H. Waddington, pp. 89–116. Chicago: Aldine. [2.4]

Thomas, R., ed. 1979. *Kinetic Logic: A Boolean Approach to the Analysis of Complex Regulatory Systems. Lecture Notes in Biomathematics*, vol. 29. Berlin: Springer-Verlag. [4.4]

Thompson, J.M.T., and Stewart, M.B. 1986. *Nonlinear Dynamics and Chaos*. Chichester, U.K.: Wiley. [A.1, A.2]

Traub, R. D., and Wong, R.K.S. 1981. Penicillin induced epileptiform activity in the hippocampal slice: A model of synchronization of CA_3 pyramidal cell bursting. *Neuroscience* 6: 223–30. [4.6]

Tulandi, T. 1985. Update in ovulation induction. *Contemp. Ob/Gyn.* 2: 1–7. [5.4]

Turek, F., and Losee-Olson, S. 1986. A benzodiazepine used in the treatment of insomnia phase-shifts the mammalian circadian clock. *Nature* 321: 167–68. [7.5, 9.4]

Tyson, J. J., and Sachsenmaier, W. 1978. Is nuclear division in *Physarum* controlled by a continuous limit cycle oscillator? *J. Theor. Biol.* 73: 723–37. [5.1, 6.2]

Van der Kloot, W., Kita, H., and Cohen, I. 1975. The timing and appearance of minature end-plate potentials. *Prog. Neurobiol.* 4: 269–326. [3.1]

van der Pol, B. 1926. On relaxation oscillations. *Phil. Mag.* 2: 978–92. [7.2]

van der Pol, B., and van der Mark, J. 1928. The heartbeat considered as a relaxation oscillation and an electrical model of the heart. *Phil. Mag.* 6: 763–75. [7.2, 9.2]

Van der Tweel, L. H., Meijler, F. L., and Van Capelle, F.J.L. 1973. Synchronization of the heart. *J. Appl. Physiol.* 34: 283–87. [7.1]

Van Meerwijk, W.P.M., de Bruin, G., Van Ginneken, A.C.G., Van Hartevelt, J., Jongsma, H. J., Kruyt, E. W., Scott, S. S., and Ypey, D. L. 1984. Phase resetting properties of cardiac pacemaker cells. *J. Gen. Physiol.* 83: 613–29. [2.3, 6.3]

Vibert, J.-F., Caille, D., and Segundo, J. P. 1981. Respiratory oscillator entrainment by periodic vagal afferents. *Biol. Cybern.* 41: 119–30. [7.1]

Waggener, T. B., Brusil, P. L., and Kronauer, R. E. 1984. Strength and cycle time of high altitude ventilatory patterns in unacclimatized humans. *J. Appl. Physiol.* 56: 576–81. [4.5]

Wallace, A. G., Sealy, W. C., Gallagher, J. J. Svenson, R. H., Strauss, H. C., and Kasell, J. 1974. Surgical correction of anomalous left ventricular pre-excitation: Wolff-Parkinson-White (type A). *Circulation* 49: 206–13. [8.2]

Walther, H. O. 1985. Dynamics of feedback systems with time lag. In *Temporal Order*, ed. L. Rensing and N. I. Jaeger, pp. 281–90. Berlin-Heidelberg-New York-Tokyo: Springer-Verlag. [4.6]

Wazewska-Czyzewska, M. 1984. *Erythrokinetics*. Foreign Scientific Publications, National Center for Scientific, Technical and Economic Information. [4.6, 9.4]

Wazewska-Czyzweska, M., and Lasota, A. 1976. Matematyczne problemy dynamiki ukkadu krwinck czerwonych. *Roczniki Polskiego Towarzystwa Mathemtycznego*, 3d ser., *Matematyka Stosowana* VI: 23–39. [4.6]

Wehr, T. A., and Goodwin, F. K., eds. 1983. *Circadian Rhythms in Psychiatry*. Pacific Grove, Calif.: The Boxwood Press. [7.5, 9.4]

Weisbrodt, N. W. 1981. Motility of the small intestine. In *Gastrointestinal Physiology*, ed. L. R. Johnson, pp. 30–37. St. Louis: C. V. Mosby. [1.1]

Weiss, R. M., Wagner, M. L., and Hoffman, B. F. 1968. Wenckebach periods of the ureter: A further note on the ubiquity of the Wenckebach phenomenon. *Invest. Urol.* 5: 462–67. [8.1]

Weitzman, E. D. 1981. Disorders of sleep and the sleep-wake cycle. In *Harrison's Principles of Internal Medicine*, 9th ed., *Update* 1, pp. 245–63. New York: McGraw-Hill. [7.5]

Welsh, B., Gomatam, J., and Burgess, A. E. 1983. Three-dimensional chemical waves in the Belousov-Zhabotinsky reaction. *Nature* 304: 611–14. [8.4]

Wenckebach, K. F. 1904. *Arrhythmia of the Heart: A Physiological and Clinical Study.* Edinburgh, U.K.: Green. [8.1]

West, B. J., and Goldberger, A. L. 1987. Physiology in fractal dimensions. *Amer. Sci.* 75: 354–65. [3.4, 9.1]

Wever, R. A. 1979. *The Circadian Systems of Man: Results of Experiments under Temporal Isolation.* New York: Springer-Verlag. [7.5]

Wiener, N., and Rosenblueth, A. 1946. The mathematical formulation of the problem of conduction of impulses in a network of connected excitable elements, specifically in cardiac muscle. *Arch. Inst. Cardiol. Mex.* 16: 205–65. [8.2, 8.3, 9.2]

Wiggers, C. J. 1940. The mechanism and nature of ventricular fibrillation. *Am. Heart J.* 20: 399–412. [8.5]

Wiggers, C. J., and Wégria, R. 1940. Ventricular fibrillation due to single, localized induction and condenser shocks applied during the vulnerable phase of ventricular systole. *Am. J. Physiol.* 128: 500–505. [8.5]

Winfree, A. T. 1972. Spiral waves of chemical activity. *Science* 175: 634–36. [1.4]

———. 1973a. Scroll-shaped waves of chemical activity in three dimensions. *Science* 181: 927–39. [8.4]

———. 1973b. Time and timelessness of biological clocks. In *Temporal Aspects of Therapeutics*, ed. J. Urquardt and F. E. Yates, pp. 35–57. New York: Plenum. [5.4]

———. 1974. Rotating solutions to reaction/diffusion equations. *SIAM-AMS Proceedings* 8, 13–31, ed. D. Cohen. Providence, R.I.: *American Mathematical Society.* [8.3]

———. 1975. Resetting biological clocks. *Physics Today* 24: 34–39. [2.3, 6.3]

———. 1977. Phase control of neural pacemakers. *Science* 197: 761–63. [5.4, 6.3]

————. 1980. *The Geometry of Biological Time.* New York: Springer-Verlag. [2.3, 6.1, 6.3, 7.3, 7.5, 8.2]

————. 1983a. Impact of a circadian clock on the timing of human sleep. *Am. J. Physiol.* 245 (*Regulatory Integrative Comp. Physiol.* 14): R497–504. [7.3, 7.5]

————. 1983b. Sudden cardiac death: A problem in topology. *Sci. Am.* 248(5): 144–60. [6.5, 8.5]

————. 1984. Exploratory data analysis: Published records of uncued human sleep-waking cycles. In *Mathematical Modelling of Circadian Systems,* ed. M. Moore-Ede and C. A. Czeisler. New York: Raven Press. [7.3, 7.5]

————. 1985. Organizing centers for chemical waves in two and three dimensions. In *Oscillations and Traveling Waves in Chemical Systems,* ed. R. J. Field and M. Burger, pp. 441–72. New York: Wiley. [8.5]

————. 1986. Benzodiazepines set the clock. *Nature* 321: 114–15. [9.4]

————. 1987a. *The Timing of Biological Clocks.* New York: W. H. Freeman. [6.1, 6.3]

————. 1987b. *When Time Breaks Down: The Three-Dimensional Dynamics of Electrochemical Waves and Cardiac Arrhythmias.* Princeton: Princeton University Press. [6.1, 6.3, 6.5, 8.2, 8.4, 8.5]

Winfree, A. T., and Strogatz, S. H. 1983a. Singular filaments organize chemical waves in three dimensions. I: Geometrically simple waves. *Physica* 8D: 35–49. [8.4]

————. 1983b. Singular filaments organize chemical waves in three dimensions. II: Twisted waves. *Physica* 9D: 65–80. [8.4]

————. 1983c. Singular filaments organize chemical waves in three dimensions. III: Knotted waves. *Physica* 9D: 333–45. [8.4]

————. 1984a. Singular filaments organize chemical waves in three dimensions. IV: Wave taxonomy. *Physica* 13D: 221–33. [8.4]

————. 1984b. Organizing centers for three-dimensional chemical waves. *Nature* 311: 611–15. [8.4]

Wingate, D. L. 1983. The small intestine. In *A Guide to Gastrointestinal Motility,* ed. J. Christensen and D. L. Wingate, pp. 128–568. Bristol, U.K.: Wright. [1.1]

Wintrobe, M. M. 1976. *Clinical Hematology.* Philadelphia: Lea & Febiger. [3.2, 4.9, 9.4]

Wirz-Justice, A. 1983. Anti-depressant drugs: Effects on the circadian system. In *Circadian Rhythms in Psychiatry,* ed. T. A. Wehr and F. K. Goodwin, pp. 235–64. Pacific Grove, Calif.: The Boxwood Press. [7.5]

Wit, A. L., and Cranefield, P. F. 1976. Triggered activity in cardiac muscle fibers of the simian mitral valve. *Circ. Res.* 38: 85–98. [5.4]

Wolf, A., Swift, J. B., Swinney, H. L., and Vastano, J. A. 1985. Determining Lyapunov numbers from a time series. *Physica* 16D: 285–317. [3.4, 4.6]

Yamashiro, S. M., Hwang, W., Sadlock, D., and Grodins, F. S. 1985. Breathing pattern regulation near apnea in anesthetized dogs (Abstract). *Federation Proceedings* 44: 1581. [5.2]

You, C. H., Chey, W. Y., Lee, K. Y., Menguy, R., and Bortoff, A. 1981. Gastric and small intestinal myoelectric dysrhythmia associated with chronic intractable nauseas and vomiting. *Ann. Int. Med.* 95: 449–51. [8.5]

Ypey, D. L., Van Meerwijk, W.P.M., and DeHaan, R. L. 1982. Synchronization of cardiac pacemaker cells by electrical coupling. In *Cardiac Rate and Rhythm*, ed. L. N. Bouman and H. J. Jongsma, pp. 363–95. The Hague: Martinus Nijhoff. [7.1, 7.4]

Zaslavsky, G. M. 1978. The simplest case of a strange attractor. *Phys. Lett.* 69A: 145–47. [7.4]

Zipes, D. P. 1979. Second-degree atrioventricular block. *Circulation* 60: 465–72. [8.1]

Zipes, D. P., and Jalife, J., eds. 1985. *Cardiac Electrophysiology and Arrhythmias*. Orlando: Grune and Stratton. [1.1, 8.1, 8.5]

Subject Index

accessory pathways, 156
acetylcholine, 36
action potential: cardiac, 60, 61, 159; neural, 59, 74; propagation, 153, 168
adaptation, 65
alternans, 145–47, 166, 167; and period doubling bifurcations, 152
amennorhea, 95
annihilation: of limit cycles, 93–95, 97, 114, 118; of sinus node oscillations, 11, 94; of traveling waves, 14, 155, 156
anovulation, 97
aplastic anemia, 181
apnea, 88, 95, 97
Arnold tongues (horns), 127, 134, 141, 200
asymptotic stability, 187
atrial fibrillation: autocorrelation function, 165, 170, 179, 180, See also fibrillation
atrial flutter, 156, 169
atrial tissue, spiral waves in, 158
attractor, 50
autocorrelation function, of atrial fibrillation, 165, 170
autonomic nervous system, 17
AV block, 124, 145, 146, 148–51, 167
AV node, 145, 156, 161

baroreceptor reflex, 136
basin of attraction, 25
basket cells, 76
Belousov-Zhabotinsky reaction, 15, 16, 157, 163, 169; scroll waves, 160, 161; spiral waves, 15, 158, 169
bifurcation: bifurcation diagram, 32; bifurcation point, 26. See also Hopf bifurcation, period doubling bifurcation, tangent bifurcation
biochemical networks, 208
birth control, and pacemaker annihilation, 95, 97
bistability, 93, 152, 153
black hole, 93, 95, 97

blood gases, during mechanical ventilation, 120
blood pressure: negative feedback control, 80; steady state, 4, 21
blood vessel rhythms, 17
bradycardia, 160
Brownian motion, 36, 41. See also random walk
Brusselator, 206
bundle of His, 145
butterfly effect, 33

Cantor function, 149, 151; Cantor set, 53
carcinoma of the bladder, 8
cardiac arrhythmias, 17, 155, 157, 175. See also alternans, atrial fibrillation, atrial flutter, AV block, bradycardia, fibrillation, modulated parasystole, multifocal atrial tachycardia, reentry, respiratory sinus arrhythmia, tachycardia, ventricular ectopic beats, ventricular fibrillation, Wenkebach block
cardiac pacemakers, 176; mathematical models for, 20, 34, 60, 61, 79, 83, 84, 96. See also SA node
cardiac rhythm, 4, 5, 10; annihilation of, 11, 94, 95, 97; phase locking, 12, 13, 121, 122, 141, 142; phase resetting, 11, 99, 101, 111, 113, 114, 115, 117
cell cycle, noise vs chaos, 44, 45, 55
cellular automata, spiral waves in, 158, 169
central pattern generators, 17, 63, 64, 79
chaos, 6, 17, 33, 35, 54; in differential equations, 62, 209; in electroencephalograms, 54, 56; in fibrillation, 165, 170, 171; in finite difference equations, 33, 199; identification of, 47–50, 54, 55, 56; in leukemia, 9, 10, 74–76, 80; in mitotic oscillators, 44, 45, 55; in mixed feedback, 72–78, 80, 81; in pancreatic